Conceptual Biomechanics and Kinesiology

Animesh Hazari · Arun G. Maiya
Taral V. Nagda

Conceptual Biomechanics and Kinesiology

Springer

Animesh Hazari
Academics and Research
Gulf Medical University
Ajman, United Arab Emirates

Arun G. Maiya
Dean, Manipal College of Health Professions
Manipal Academy of Higher Education
Manipal, India

Taral V. Nagda
Pediatric Orthopaedic Surgeon
SRCC Children Hospital
Director Jupiter Gait Lab
Mumbai, India

ISBN 978-981-16-4990-5 ISBN 978-981-16-4991-2 (eBook)
https://doi.org/10.1007/978-981-16-4991-2

This Springer imprint is published by the registered company Springer Nature Singapore Pte Ltd.
The registered company address is: 152 Beach Road, #21-01/04 Gateway East, Singapore 189721, Singapore

Foreword 1

"Conceptual Biomechanics: A Comprehensive Textbook" The biomechanics and kinesiology applied to human body is quite complex though interesting. For students coming from a biology background, the subjects are quite challenging as they hold in-depth understanding of physical and mathematical principles. Therefore, it requires to be delivered in a conceptual manner. It would then be beneficial for easy understanding of the human movement science and applications of forces on human body.

I am delighted to write the foreword, to this proactive monograph, which sheds light on a different approach to biomechanics of human body. I compliment the authors Dr. Animesh Hazari, Dr. Arun Maiya, and Dr. Taral Nagda for conceptualizing this book. The book has been written in very lucid and simple language. The contents of the book are current, up to date, and well researched. I once again congratulate the authors for their committed efforts, and wish them all the very best. I am sure that this monogram will be a welcome addition as a very useful resource to all students and professionals. This book should find a place in the libraries of all Medical and Allied Health Institutions.

H. S. Ballal
Pro Chancellor, Manipal Academy of Higher Education
Manipal, Karnataka, India

Foreword 2

It gives me great pleasure to write the foreword to the book titled "*Conceptual Biomechanics and Kinesiology*". The biomechanics is an advanced science which has driven our ability to objectively describe human movements in all means.

The Manipal Academy of Higher Education has always promoted and encouraged our faculty members to engage in research on diseases of critical importance in a global perspective and come up with newer evidence-based solutions and treatment options for the benefit of large global community.

I am delighted to write the foreword, to this proactive monograph, which sheds light on a different approach to clinical biomechanics and kinesiology. I compliment the authors Dr Animesh Hazari, Dr Arun Maiya and Dr Taral Nagda for conceptualising this book. Dr Arun Maiya whom I have known for long is a senior advisor to Indian Podiatry Association (IPA) and President of Karnataka Chapter of IPA and leading the Centre for Diabetic Foot Care & Research at MAHE, Manipal.

The book has been written in a very lucid and simple language. The contents of the book are current, up to date and well researched. The chapters have been very well structured with illustrations used very judiciously and imaginatively to hold the interest and to emphasise important points for easy understanding and retention.

I am sure that lots of students, Doctors and other professionals who are actively involved in healthcare would be immensely benefited with this book.

I once again congratulate the authors for their committed efforts towards authoring a quality book and wish the authors all the very best.

Lt. Gen. (Dr) M.D. Venkatesh
Vice Chancellor, Manipal Academy of Higher Education
Manipal, Karnataka, India

Foreword 3

Biomechanics and Kinesiology are important principles that are required to understand the functioning of the human body from a rehabilitation point of view.

Physical therapy students are required to understand the functioning of human joints by applying the knowledge of Biomechanics and Kinesiology to normal anatomy, to compare and evaluate the joints to understand the problems, and to monitor the progress after rehabilitation interventions.

This book gives a deep insight to understand and apply principles of Biomechanics and Kinesiology on all peripheral and spinal joints. Emphasis is also placed on Gait analysis and interpretation of gait parameters.

Application of different principles of Biomechanics and Kinesiology and utilization of the same information in evaluation, treatment, and monitoring the progress of patients is very crucial in successful practice of Physical therapy.

This book will be of use for students, teachers, and practicing physical therapists, and professionals of Sports Sciences and Health education experts.

I appreciate the efforts of Dr. Animesh who is an expert and experienced in this field who has authored this book to make these essential principles very easy and understandable in this book.

Praveen Kumar
Dean, College of Health Sciences
Gulf Medical University
Ajman, UAE

Preface

I dedicate this book to all my students who believed in my teaching skills. Their constant feedback has laid the path for preparing this book. Being a student myself, I found that there was always a need to conceptualize the human movement science for easy understanding. I also believed that students at the bachelor level need to correlate the biomechanical principles to clinical practice and the effort has been made to fulfill this objective. As the textbook, this book would provide essential linkage between the basic and advanced concepts of Biomechanics as well as Kinesiology. The book has been prepared keeping larger disciplines of students including but not limited to medicine, allied health, and biomedical graduates who will benefit themselves through easy conceptualization of the subject matter.

This work is also part of my constant effort to benefit the society and give back the knowledge that I have gained in my academia. The path would not have been easy without the support of my parents and family members (Mr. Krishna Kumar Hazari, Mrs. Rita Hazari, Mr. Vijay Thakur, and Mrs. Arpita Thakur), in-laws (Mr. Sanjay Jha, Mrs. Renu Jha, Mr. Kuldeep Jha, and Mrs. Priyanka Deshpande), wife (Mrs. Rakhi Hazari), and beloved daughter (Anika Hazari). At last, I would like to thank all my supervisors, mentors, colleagues, and students who have contributed in shaping my abilities and potential at this platform. My sincere thanks to Dr. Hafiz Wani for his support and time as a proof reader.

Ajman, UAE — Animesh Hazari

The original version of this book was revised: The "Foreword" page has been included in the front matter.

The original version of this has been updated: Second Author Arun G. Maiya affiliation has been updated. A Correction to this book is available at https://doi.org/10.1007/978-981-16-4991-2_21

Contents

List of Figures

List of Tables

Concepts of Biomechanics

1

1.1 Introduction to Biomechanics

Locomotion is the inherent ability of any living organism. Have we ever thought what is the science behind a bird flying, a human walking, a reptile crawling, a panther running, or any movement that we can capture through the naked eye or microscopically? It is the concepts of biomechanics that deal with living body movement patterns and their driving force. The term "biomechanics" could be split as "bio"-living body (in particular to humans as we will discuss in this book) and "mechanics"-movement science. Therefore, it is defined as the study of the movement of living things using the science of mechanics [1]. In other terms, the branch of mechanics that deals with the movement of the human body would be referred to as biomechanics in this book. For example, if we study the science behind our walking, running, swimming, etc., we are dealing with biomechanics.

1.1.1 Need for Biomechanics

(a) Efficient movement—The biomechanical concepts help us to learn the most efficient patterns of movements in terms of maximum output with minimum effects. For example, you may have learnt that if you throw a javelin at 45 degrees the distance covered would be maximum. This is a universal physical principle, but what about if your joint and muscles mechanics are incorrect. Would you be able to reach the desired distance? Thus, it is the study of biomechanics that allows for an efficient movement pattern of the human body.
(b) Improved performance in sports—Sports biomechanics is one of the most advanced areas where the concepts of biomechanics are used to improve performance in sports, for example, technique analysis and correction using motion devices and training on biomechanical devices to obtain the desired output.

A. Hazari et al., *Conceptual Biomechanics and Kinesiology*,
https://doi.org/10.1007/978-981-16-4991-2_1

(c) Clinical training and rehabilitation—The biomechanical concepts have been widely used for rehabilitation and clinical training for recovery from injuries and pathomechanics. For example, it is the principles of biomechanics under which a footballer is rehabilitated for ligament injuries or a rotator cuff is managed.

The dimensions of biomechanics as the subject are well spread because the complexity of movement varies a lot. However through the well-established concepts, we shall try to learn about the human biomechanics pertaining to its properties of motion and force at large. We shall now understand the subdivision of biomechanics which would help us to understand the subject better and build a stronger base for the upcoming sections. The biomechanics has two major branches to deal with, viz. **kinematics and kinetics**.

1.1.2 Kinematics

The branch of biomechanics that deals with pure motion of the body without considering the forces acting on it can be referred to as kinematics. For example, let us assume a human walking (Fig. 1.1).

All the parameters that describe motion would be kinematics. Let us understand in depth with clarity. Consider the above picture which is describing a human walk. Now imagine yourself doing a normal walk at your home. You walk with a comfortable **speed**. In another instance, when you are getting late to college, you are rushing with more **speed**. In both cases, you are walking and considered to be in a state of motion, and the parameter that describes the motion is **speed** which is changing as per the circumstances. Therefore, speed is one of your kinematics.

Let us have further insights to some more kinematic variables. Now consider the first instance where you were walking at a slower and comfortable speed, your number of steps and **step length** would be smaller, but when you walk faster, your **step length** would be larger to cover more distance in shorter time. Therefore, step length is again a kinematic variable which is describing the motion.

Similarly, when you walk you also notice that your joints keep changing their angle of motion; thus, **angle** is again a kinematic variable which signifies motion (Fig. 1.2).

1.1.2.1 Osteokinematics

The branch of kinematics dealing with the movement of bones, e.g., upward and downward displacement of Humerus when we do shrugging.

1.1.2.2 Arthrokinematics

The branch of kinematics dealing with the movement of joints, e.g., translation or rotation of the glenohumeral joint when we do flexion, extension, abduction, etc.

In order to understand the biomechanics of joint functions, the concept of osteo- and arthrokinematics is very important and the same principles would be used throughout this book.

Fig. 1.1 Depicts kinematics of human walking

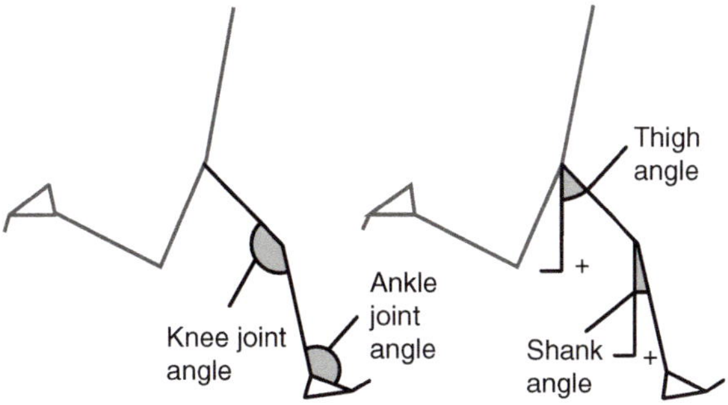

Fig. 1.2 Describing joint angles (kinematic variable)

1.1.3 Components Describing Kinematics

The major components that describe kinematics are:

1. Nature and type of motion
2. The point of motion (location)

3. The direction and magnitude of motion
4. The rate of change of motion

Let us understand all these above in detail.

1.1.3.1 Nature and Type of Motion

Translatory Motion: a type of motion or displacement that occurs in the straight line and thus also known as the **linear displacement**. It is important to understand that a translatory motion takes place in a rigid segment where all points on the given segments move through equal distance, in the same straight line, parallel, and in the same point of time. However, in the human body a true translatory motion is an assumption as the human body has linked joint segments that move independent to each other but not as a single segment. The best example for a translatory motion for human system could be given as movement of the humerus upward and downward while doing shrugging (Fig. 1.3). Assuming that bones of the forearm and arm act as a single segment we can see that the humerus translates upward as we do shoulder shrugs and translate downward as we relax. It should be also noted that all translatory motion should be considered in the plane of the human body or ground as a reference.

Rotatory Motion: a type of motion or displacement that occurs in curved path using a fixed axis/center of rotation. For a true rotatory motion all points on the segment should move through the same angle and at a constant distance from the axis at the same time. For a better understanding, let us consider that you started a wheel

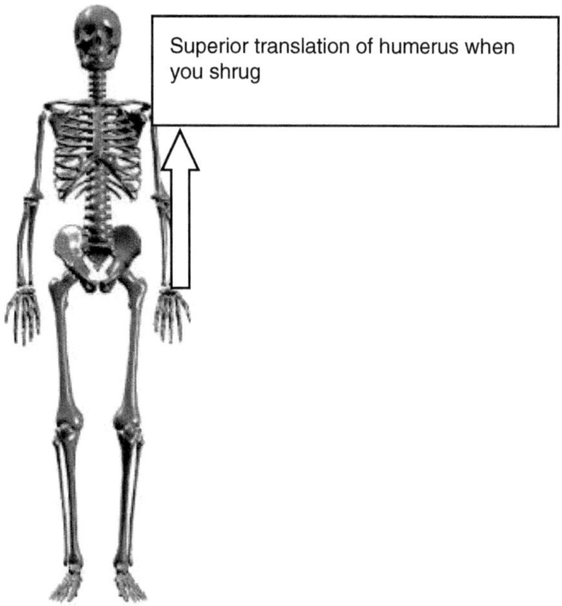

Fig. 1.3 Depicting translation of the humerus at the shoulder joint

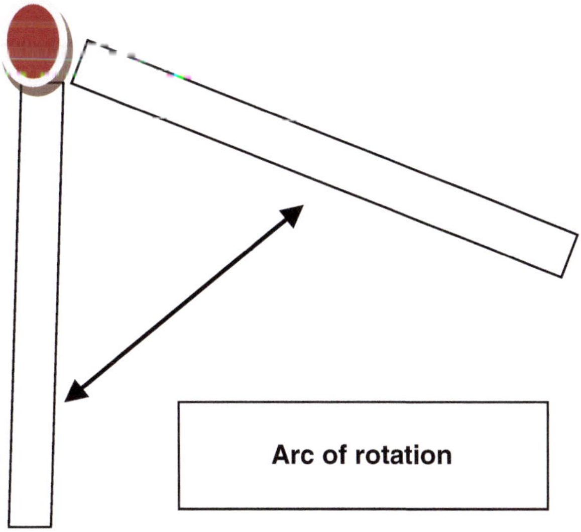

Fig. 1.4 Depicts rotatory motion of the right shoulder while doing abduction

to rotate with your hand at a constant speed. At each point of time frame, the wheel will rotate with the same angle over the same fixed axis. Since the motion is in angles, it is also known as **angular displacement.**

Unfortunately, in the human body no true rotatory motion occurs as the body segment keeps changing the axis constantly. Thus, the human body could be considered as complex machine which have the ability to change its axis constantly even for rotatory motions, making it smoother and not mechanical as seen with shoulder abduction movement (Fig. 1.4).

Combined Motion: In the human body, the motion that predominantly occurs is the combination of translatory and rotatory, thus making it biomechanically most efficient. When the combined motion occurs in the two dimensions, it is known as curvilinear motion or planar motion where the axis of rotation keeps shifting around an instantaneous center of rotation. For example, flexion and extension of knee has the combined motion of translation and rotation of the tibia over the femur which occurs in three dimensions over an instantaneous center of rotation (Fig. 1.5). In the human body where the motion occurs in three dimensions, the combined motion takes place around a helical axis or screw axis, e.g., shoulder circumduction. This principle is applicable to all joints of the human body, and it would be important to understand the concept for further correlations in upcoming chapters.

1.1.3.2 The Point of Motion

In the human body, the reference to the origin of motion is taken from the point known as the center of mass (COM) where the entire mass of the human body is assumed to be concentrated. Since the human body is considered a free body, each segment has its own COM. From this point, the Cartesian coordinate system of axes and planes is drawn to understand the human kinematics. As there are three axes and three planes around which a human body motion occurs, it is also referred to as

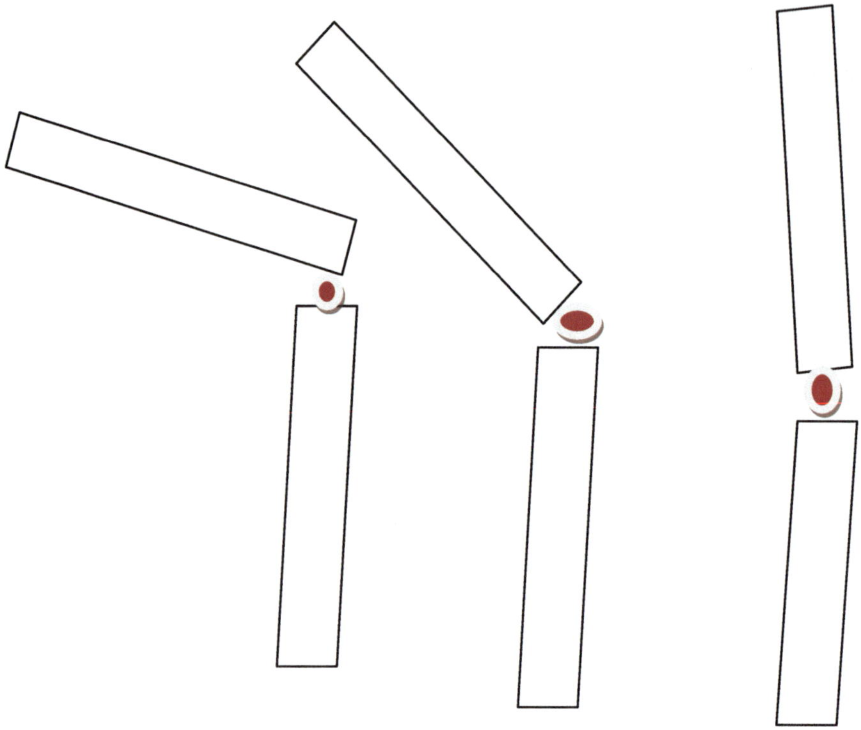

Fig. 1.5 Showing combined motion for knee flexion over an instantaneous axis. It can be seen that the center of rotation keeps shifting instantaneously from full knee flexion to extension

three-dimensional motions in the x-axis, y-axis, and z-axis. In context to the human body and the anatomical position as shown in Fig. 1.6, the x-axis runs side to side and is known as the **coronal/ frontal axis**. Similarly, the y-axis runs top to bottom of the body and thus is known as the **vertical/ longitudinal axis,** whereas the z-axis runs from front to back of the body which is known as the **anteroposterior/sagittal axis.** Accordingly any motion taking place through the **frontal axis** would be considered in the **sagittal plane**, whereas motion occurring through the **vertical/longitudinal axis** and **anteroposterior/ sagittal axis** would be taking place in the **horizontal/transverse plane and frontal plane,** respectively, as shown in Fig. 1.6. The motions occurring in axes and planes form **degrees of freedom.** For example, the shoulder joint has motions available in all three axes and three planes; thus, it has got six degrees of freedom.

In context to the human body, when you cut the human body into two halves from front and back, you obtain the frontal section (plane as referred here). When you divide the human body into two halves from medial and lateral, you obtain the sagittal section, and when you cut the body through the up and down, you obtain the transverse section. This applies to all human body segments cut at any point.

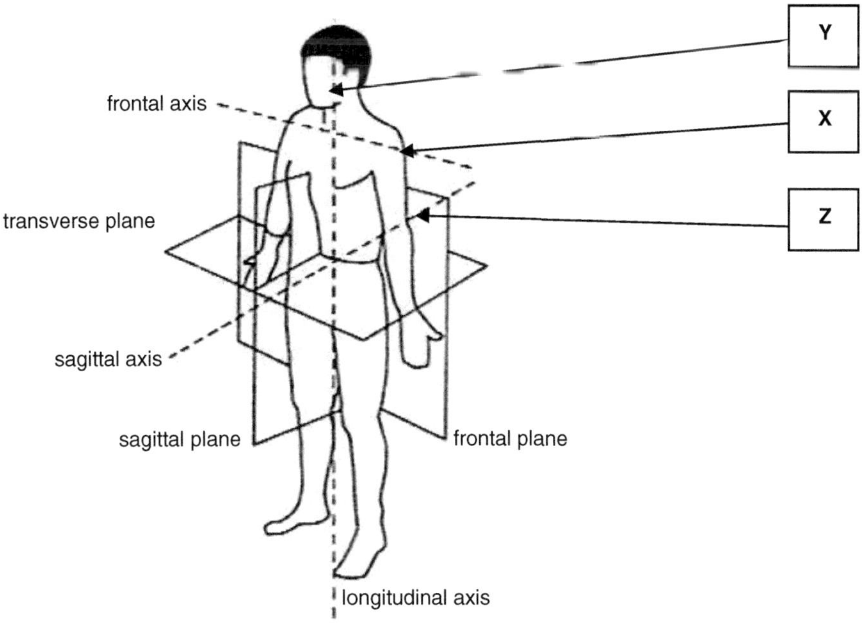

Fig. 1.6 Depicts axes and planes for human kinematics

Table 1.1 Anatomical description for human segment motions in axes and planes

Axis	Plane	Predominant motion at major joints[a]
Frontal axis	Sagittal plane	Flexion/extension of neck, shoulder, elbow, wrist, spine, hip, and knee Ankle dorsiflexion and plantar flexion
Sagittal axis/ anteroposterior axis	Frontal plane	Neck side flexion to left and right Shoulder adduction and abduction Wrist radial and ulnar deviation Spine side flexion to right and left Hip adduction and abduction Ankle supination and pronation
Vertical axis	Horizontal/ transverse plane	Neck rotation to left and right Shoulder medial and lateral rotation Elbow supination and pronation Hip medial lateral rotation Ankle inversion and eversion

[a]Only predominantly seen motions at various joints have been given as example here

Since you may have learnt various motions of the human body, let us correlate them with axis and plane motions as shown in Table 1.1. To make you understand the concepts of axes and planes in more detail, let us follow the experiment below.

Steps to follow:

Take a notebook cover or cardboard and assume it to be the plane.

Take a stick/pen and consider it to be the axis.

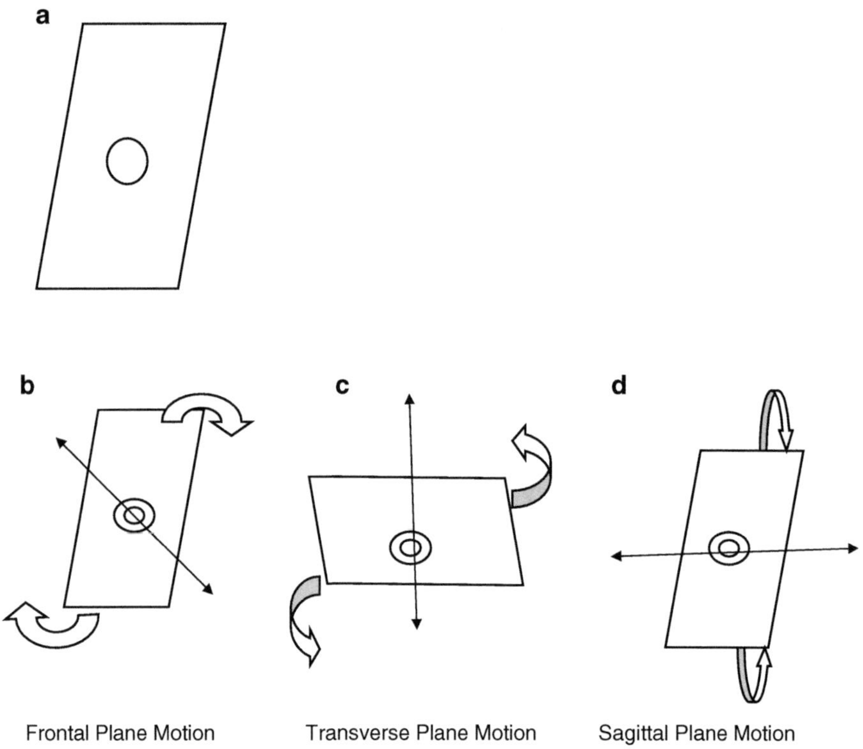

Fig. 1.7 (**a**) Showing a cardboard with a center at COM. (**b**) Frontal plane motion. (**c**) Transverse plane motion. (**d**) Sagittal plane motion

Now, mark a center point in the notebook cover or cardboard and make a small hole as shown in Fig. 1.7a.

Insert the pen through the hole.

Now understand the axis and plane kinematics with placement of notebook and pen in three different positions as shown in Fig. 1.7b, c, and d.

In Fig. 1.7b, the stick (axis) is passing from front (anterior) to back (posterior) through the notebook (plane). Now keeping the axis fixed, let us rotate the notebook. The only direction that we can rotate the notebook is left and right (see Fig. 1.7b). This movement can be referred to as abduction and adduction in the human body like shoulder joint (Table 1.1) where the axis is passing anteroposterior and plane of movement is frontal.

In Fig. 1.7c, the stick (axis) is passing from top to bottom through the notebook (plane). Now keeping the axis fixed, let us rotate the notebook. The only direction that we can rotate the notebook is front and back (see Fig. 1.7c). This movement can be referred to as medial and lateral rotation in the human body like spine (Table 1.1) where the axis is passing vertically and plane of movement is transverse.

In Fig. 1.7d, the stick (axis) is passing from left to right or medial to lateral through the notebook (plane). Now keeping the axis fixed, let us rotate the notebook. The only direction that we can rotate the notebook is anterior and posterior (see Fig. 1.7d). This movement can be referred to as flexion and extension in the human body like the shoulder, elbow, hip, knee, and spine (Table 1.1) where the axis is passing frontal and plane of movement is sagittal.

1.1.3.3 The Direction and Magnitude of Motion

In order to understand the direction and magnitude of human motion, you must first know the scalar and vector physical quantities. A scalar is a quantity that has only magnitude but no direction, whereas a vector quantity has both direction and magnitude. For example, consider that you are running. One of the kinematic variables that describe your motion (running) is the speed or velocity. We can say that you are running at a "speed" of 20 km/hr. which is scalar, but when you also add direction to it, the same quantity becomes vector that is "velocity." So now you can say that you are running at a velocity of 20 km/hr. in the east direction. Similarly, mass and weight, distance and displacement, etc., are examples of scalars and vectors.

It is important to understand that the direction of vectors is denoted by an arrow and the length of the arrow determines the magnitude of the vector. In the example below, we can see that a vector variable like displacement shows the direction of movement denoted by arrow and the length is maximum for the highest magnitude.

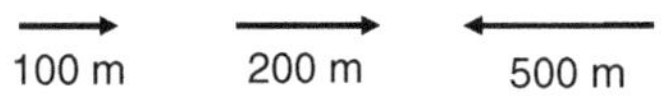

1.1.3.4 The Rate of Change of Motion

For understanding kinematics, the rate of change of motion is also very important. The most common kinematic variables include joint angles, velocity, speed, angular acceleration, angular velocity, momentum, etc. The three-dimensional motion analysis systems can detect the rate of change precisely and thus are very helpful in determining the clinical and sports kinematics.

1.1.4 Most Common Kinematic Variables for Human Biomechanics

(a) Speed /velocity—rate of change of distance upon time
(b) Distance—total covered area
(c) Displacement—difference between the starting and end point
(d) Acceleration—rate of change of velocity in respect of time
(e) Joint angles—angles formed between the adjacent bones
(f) Spatiotemporal gait parameters like step length, step time, stride length, stride time, and toe angle

I believe that kinematics is well understood now. We shall now move to kinetics in the next section.

1.1.5 Kinetics

The branch of biomechanics that deals with all internal and external forces causing a motion is referred as kinetics. Let us understand kinetics with Fig. 1.1 which describes a human walk. Now when you walk, all factors that are responsible for us to keep in motion (walking here) would be kinetics. For example, it is the muscular force that allows us to walk; there is frictional force between the joints and there is weight of your body as force acting downward and ground reaction force exerting upward (Fig. 1.8). Thus, it is the combination of internal and external forces that act on a human body where the resultant force leads to dynamic and static mobility as well as stability.

1.1.5.1 Defining Force

A force is a vector quantity obtained as the product of mass and acceleration [2], i.e., $F = m \times a$, where

F force in Newton (N)
m mass of the body in kilogram (kg)
a acceleration of the body in meter/per second square (m/s^2)

1.1.5.2 Internal Force

The forces produced within and by the body itself can be considered as the internal force. For example, force produced by muscles to move the joint (pull of quadriceps muscle produces extension at knee joint), force of ligaments and tendon while contraction of muscles, and joints force between two adjacent bones (shear force).

1.1.5.3 External Force

The forces acting on the body but produced by the external surroundings and environment are considered as external forces, e.g., force of gravity and frictional force.

1.1.5.4 Resultant Force

At a given point, there can be more than one force acting on the body. The direction as well as the magnitude of the movement is then decided by the resultant force. Let

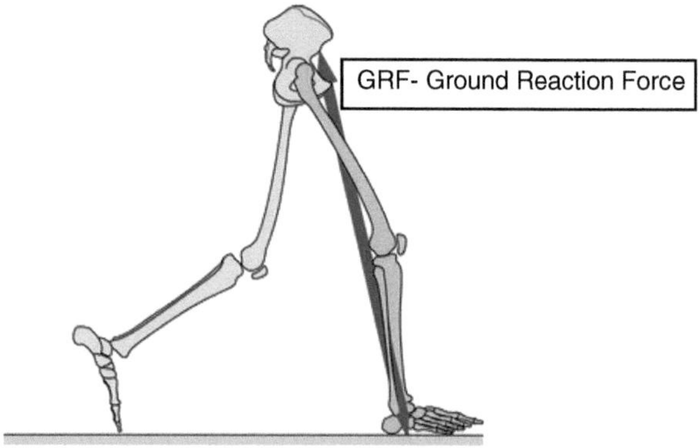

Fig. 1.8 Depicts ground reaction force (kinetic variable)

us understand this with the example below; a body experiences a force of 10 N in the east direction. If we apply another force of 15 N in the west direction, in which direction the body should move?

Since the force is a vector quantity we can do mathematical operations and determine the resultant direction and magnitude. In the given example calculate as follows:

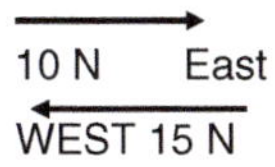

It clearly shows that body should finally move in the west direction with 5 N of resultant force.

In the human body, this concept can be applied to linear motions. However when rotations are involved, as frequently seen at human joints, where two perpendicular forces act on the body, the resultant force is determined by the triangular law or the parallelogram law. Let us understand this in more detail.

Consider two forces acting as shown by vector representation below:

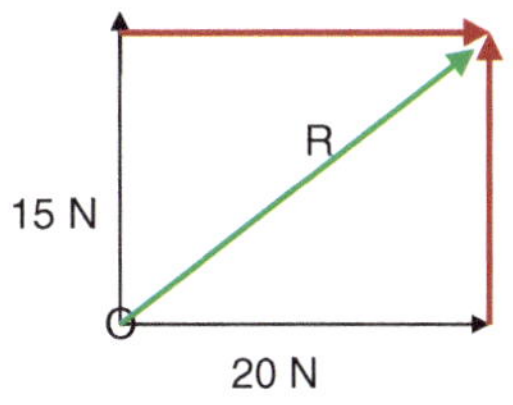

Point O is the origin of forces where a 15 N force is acting upward and another force of 20 N is acting perpendicular to it (at 90 °). For calculating the resultant force, complete the parallelogram as shown with colored red lines. The resultant force would be acting in the direction of the diagonal of the parallelogram. The magnitude can be calculated using the laws of sines and cosines which can be understood well in the kinesiology section of this book. For a right-angled triangle, Pythagoras theorem can be used to compute the resultant force as shown below.

$$R^2 = 15^2 + 20^2$$
$$R^2 = 225 + 400$$
$$R = \sqrt{625}$$
$$R = 25\,\mathrm{N}$$

Note: The resultant force is considered to be acting on a single body unit. However in the human body, the joints are made up of different bony segments; thus, the resultant forces should be considered acting on each segment differently (discussed under the kinesiology section).

1.1.5.5 Resultant Forces in a Linear Force System

In the linear force system, all the forces act on the same plane in the same parallel direction. Therefore, the resultant force would be the arithmetic sum of the forces in the direction of the acting forces. All translatory forces on the human body are part of the linear force system. For example, the force of gravity pulls the shoulder downward, but the force of deltoid muscle acts in the upward direction to keep it upright where the net or the resultant force is also in the upward direction which does not allow it to sag down (Fig. 1.9). In human kinematics, the linear force systems are mainly applicable to the static stabilizers.

Types of Forces in Linear System

(a) **Force of Gravity**

The force of gravity, also known as the gravitational force, is one of the most evident force that every physical body experiences. The force of gravity is a vector quantity that acts on individual segments of the body at the point of application known as center of mass (COM) or the center of gravity (COG) [2]. The COM and COG is the point at which all mass/weight of the body is considered to be

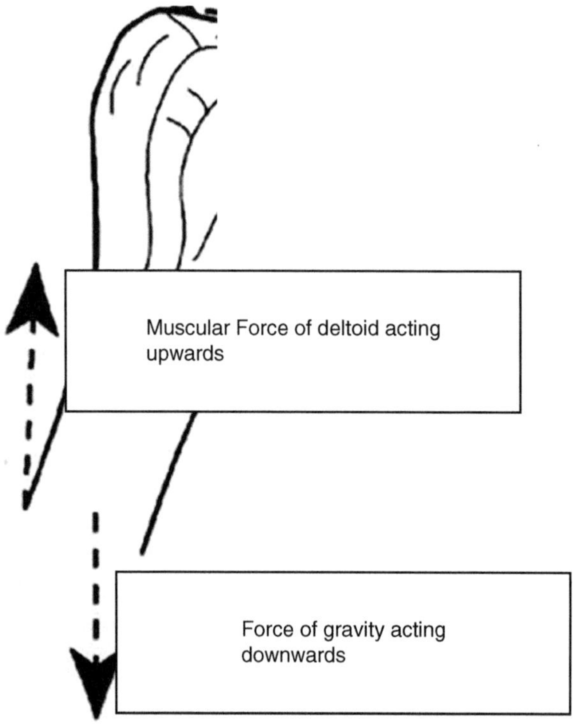

Fig. 1.9 Depicts the deltoid muscle force against gravity as part of linear force system

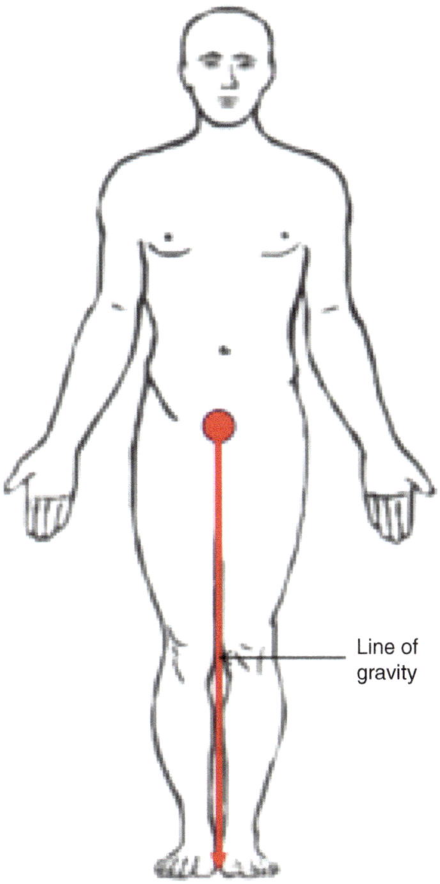

Fig. 1.10 Depicts the line of gravity (LOG) in the human body

concentrated. The center of mass is often located at the heavier part of the body and tends to shift toward the added load part. The force vector (gravitational force) starting from this point constitutes the line of gravity (LOG). For the human body, the COM lies at the vicinity of S2 vertebra. The LOG is shown in Fig. 1.10.

Application: Static and Dynamic Equilibrium

When the body is at rest, it is said to be in the static state while it is moving it attains the dynamic state. The equilibrium is a state of mechanics at which the body is stable, and COM, COG, as well as LOG all fall within the base of support (BOS) [2]. In the static state the equilibrium is well maintained. But when a body is moving, the COM or COG shifts to maintain the stability. For instance, when the COM and COG falls outside, the body becomes unstable and falls.

Experiment:

Stand on two legs with feet apart (Fig. 1.11a). You are quite stable as the BOS is larger.

Now stand on one leg (Fig. 1.11b).

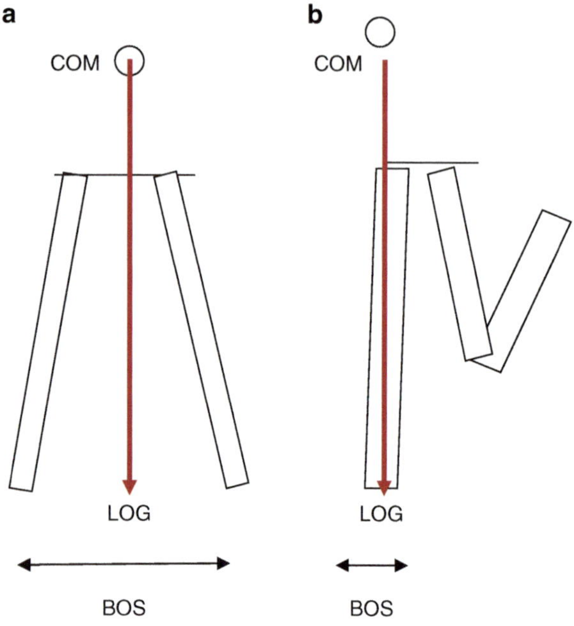

Fig. 1.11 (**a**) Depicts LOG within larger BOS, standing on both legs, (**b**) shifting of LOG within smaller BOS, standing on single leg

If we look at double leg standing, the COM and COG lies at the same point where the BOS is larger; thus, the body is more stable. When you stand on one leg, the COM shifts toward the heavier side (the weight bearing leg); thus, the LOG also shifts on a smaller base of support which makes the body less stable compared to double leg standing. It should be noted that 'the closer the COM, and LOG to BOS, the higher would be the stability'. Therefore, animals with four legs are more stable because their base of support is larger as well as the center of mass is lower compared to humans (discussed in more detail under the Posture, Chap. 13). For a punching bag, the COM, LOG, and BOS all lie at the same point which makes it most stable. Even if you thrash it to ground, it pulls itself up to upright position toward the COG. This is the universal physical law of static and dynamic equilibrium. This concept would help you to understand the biomechanics of the human body much better.

(b) **Tensile Force**

Two apposite pulls create a tension in the structure which is known as tensile force. For example, if two friends pull a heavy object through a rope toward each other, there is tension in the rope with two force vectors in the opposite direction as shown in Fig. 1.12.

Application: Stretching of Muscles
While stretching our muscles we apply a sustainable force while pulling the muscles away from its attachment which creates a tensile force within the muscle fibers.

(c) **Compressive Force**

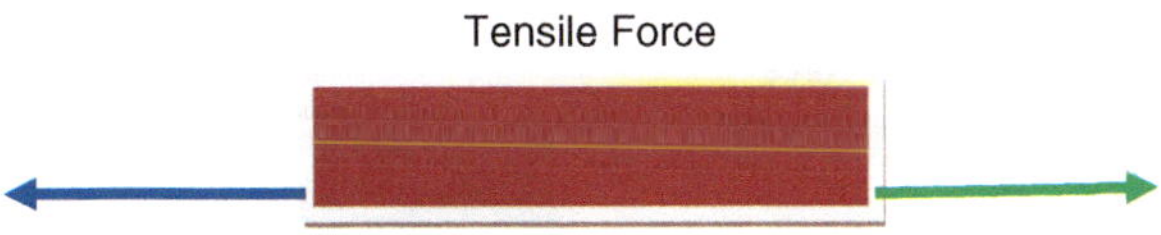

Fig. 1.12 Depicts tensile force created by two forces acting in opposite directions

Fig. 1.13 Depicts compressive force created by two forces acting in the same direction

Two forces acting in the same direction create a compression as shown in Fig. 1.13. In context to human joints it can be referred to as joint reaction force.

Application: Bone Mineralization, Synovial Fluid Diffusion
After an episode of fracture and callus formation, we try to load the bones and start walking. The body weight and the ground reaction force create a resultant compressive force which helps in calcification. Similar principle applies when the vertebral disk is compressed and synovial fluid is released.

(d) **Shear Force**

A shear force is a perpendicular force applied to the part of a surface in the direction of the movement (Fig. 1.14).

Application: Joint Mobilization and Manipulation
During mobilization a shear force is given through joint play movements in the direction of the desired movement.

(e) **Distraction Force**

In the human body, distraction force is quite common where two surfaces pull apart. This is possible when you pull even one segment, keeping the other stable or you can pull both segments apart which causes distraction. For example, you can hold the humerus with one hand and radius with the other and pull them apart or hold the humerus stable and pull radius solely. It should be noted that distraction forces in the human body are seen at the distal most segment. This force helps the segment to mobilize which is a very efficient technique in physical therapy.

Fig. 1.14 Depicts shear force created by force acting in the direction of movement

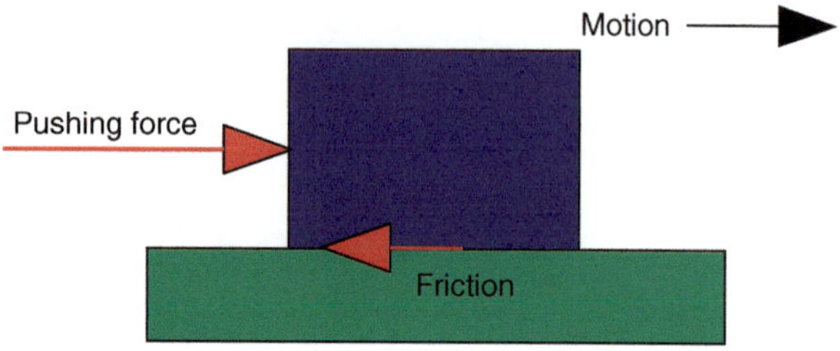

Fig. 1.15 Depicts friction force at the surface, created by forces acting in opposite directions

Application: Traction Techniques for Adhesion Release

(f) **Frictional force**

The force of friction is referred to as the opposite forces between the two surfaces (Fig. 1.15). The frictional force is a restricting force that limits the motion.

Application: Walking without slipping

1.1.5.6 Resultant Forces in a Concurrent Force System

In the concurrent force system, where multiple forces act on a given point or body segment, but in different planes and in different directions, the resultant force is determined by using the law of parallelogram as discussed above. In human kinematics, the concurrent force system prevails as the resultant of rotatory and translatory motions. The internal forces caused by muscles and soft tissues commonly form the part of concurrent force system.

Types of concurrent force system:

(a) Torque

When two forces are applied to the same object at two different points, it creates a rotational force. The resultant force is known as the torque or moment of force. Ideally, the two forces form the force couple that creates a magnitude of torque as the product of force and perpendicular distance (moment arm) from the point of application. It is measured in Nm (Newton meter).

$$T = F \times d$$

Application: Force generated by muscles. In the human body, the torque produced by muscles depends upon the position and angle of the joint. This is the reason you can lift heavy weights easily when your elbow is flexed 120 degrees to 90 degrees as the moment arm is longer and torque produced is higher compared to full extension and near flexion. See Fig. 1.16.

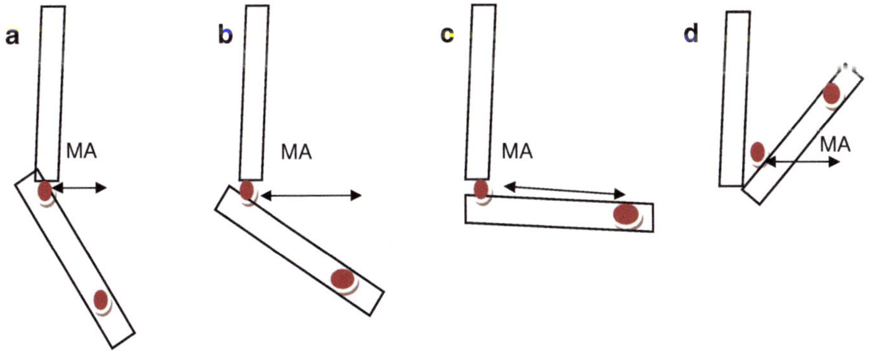

Fig. 1.16 Depicts moment arm (MA) at different positions of elbow joint. Note that MA is the longest between figure **b** and **c**

For figure b and c, the moment arm is larger; thus, the torque will also be higher in magnitude, and thus, the muscle can produce more force compared to position a and d.

(b) **External moment and internal moment**—The torque or moment of force produced as the resultant of force of gravity is known as the external moment. The direction of external moment depends upon the position of line of gravity (LOG) from the joint center as explained using Fig. 1.17.

In human biomechanics, the role of gravity is inevitable which creates a significant external moment, and all muscles work to counter the external moment to maintain the state of equilibrium which we may not realize with the naked eye. Let us understand this with the figure above where the LOG (dot line) passes through the ankle, knee, and hip joint differently. In Fig. a, the LOG passes anterior to the ankle (sagittal plane view), anterior to the knee, and anterior to the hip joint. Now the LOG would create a torque equal to product of ground reaction force (GRF acting against body weight) and moment arm depending upon the distance from the joint center. As a result in Fig. 1.17a, maximum external moment would be created at the knee and least in Figure c.

The other question is what would be the direction of this moment?

As a simple rule, keep the distal segment stable and move the proximal segment toward the LOG; the motion caused would be the direction of the external moment. Here at the ankle, the tibia is the proximal segment; thus, it would move toward the LOG causing a lesser angle between the ankle joint. Thus, we can say that the gravity has caused a dorsiflexion external moment at the ankle in Fig. a. In order to counter it and maintain stability, the muscles should also generate force which should be apposite and equal to the external moment. Thus, our plantarflexors generate a flexion **internal moment** against the external moment created by gravity. Similarly an external extension moment at the knee can be seen (Fig. 1.17a and b). Since the LOG passes directly from the knee joint center in figure c, there would be

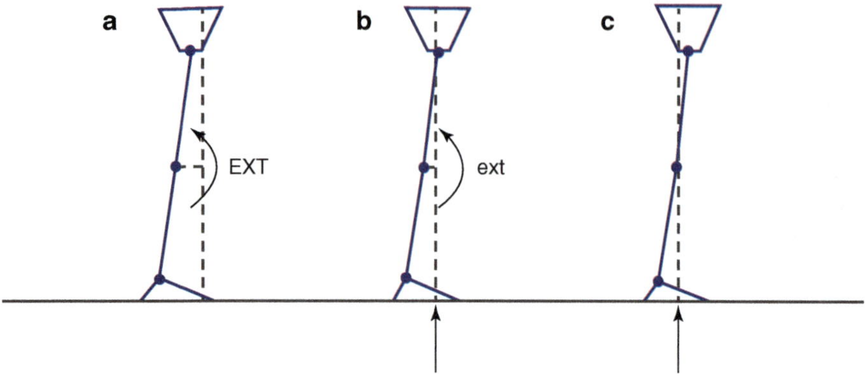

Fig. 1.17 Depicts the external moment due to LOG (↑) and its magnitude and direction

no external moment created. The external moment at the knee joint should be counteracted by the flexion internal moment of the hamstring.

1.1.6 Most Common Kinetic Variables for Human Biomechanics

(a) Joint force—Internal force from the muscles to maintain stability and mobility of joints.
(b) Ground reaction force—Force exerted in reaction to body weight in the upward direction toward the joint and body.
(c) Impulse—Force per unit time.
(d) Torque—A force component to produce rotatory effects on the joints.
(e) Power—The product of force and velocity.

1.2 Summary

The study of biomechanics in the human body comprises kinematics and kinetics. Kinematics describes the motion without considering forces, whereas kinetics deals with forces. In the human body, the linear and concurrent force systems are seen which are resolved to give the resultant force.

References

1. Hatze H. The meaning of the term "biomechanics". J Biomech. 1974;7(2):189–90.
2. Levangie PK, Norkin CC. Joint structure and function: a comprehensive analysis. Philadelphia, PA: FA Davis Co; 2011.

Cellular Biomechanics of Soft Tissues

2

2.1 Background

In this chapter, we shall focus on the composition, structure, and functions of soft tissues at the cellular level. This is important to understand the behavioral difference for kinetics and kinematic parameters among various tissue types. This can also help us to determine the morphological difference and biomechanical properties at large.

2.1.1 Classification of Soft Tissues

The soft tissues are specialized mesenchymal tissues which support the cells and organs. The soft tissues present in the human body originate from the non-epithelium basically known as the mesenchymal cells. The most important classes of soft tissues include fibrous tissue, fat, muscles, synovial tissue, lymph vessels, blood vessels, and nerves.

2.1.2 Connective Soft Tissue

In scope of the given chapter, connective soft tissues are of utmost importance to us that forms the basis for biomechanical structure and functions of the human body. The important connective tissues have been explained below.

Fibrous Tissue
The most commonly found connective tissues that are composed of thread-like structures are known as the fibers. The main function of the fibrous tissue is to

A. Hazari et al., *Conceptual Biomechanics and Kinesiology*,
https://doi.org/10.1007/978-981-16-4991-2_2

provide strength and stability which is most commonly found in muscles and tendons [1].

Muscle

Muscles are specialized connective tissues that connect bone to bone. Three main types of muscle tissues include smooth, skeletal, and cardiac.

Smooth muscle is involuntary in function such as found in the walls of the stomach, intestine, bladder, and blood vessels. Cardiac muscle forms the walls of the heart and allows the heart to pump blood. Skeletal muscles function voluntarily to control human movements such as muscles of the arm, legs, and back. In this book, our main focus would be on the skeletal muscles.

Tendon

Tendons are connective tissues that connect muscles to bone and help transmit forces originating from muscle to bone in human biomechanics, e.g., quadriceps tendon and Achilles tendon.

Ligaments

Ligaments are soft tissues that connect one bone to another or may be attached to a joint capsule to provide stability to the joints. A number of ligaments are present in the human body that act as secondary stabilizers to the joint.

Bone

These are the hardest connective tissues forming joint of the body. It helps to take stress and forces imposed on it while allowing the movement of the human body under static and dynamic positions.

Synovial Tissue

Synovial tissue is a thin and loose connective tissue that lines joints, such as the elbows and knees. The main function of the synovial tissue is to provide lubrication to the joints for smooth movement.

2.1.3 Important Cells of Soft Tissues

Chondrocytes are the main cell type in cartilage [1]. Cartilage, similar to bone, is a specialized form of connective tissue.

Fibroblasts are an important cell type that produces collagen fibers which are the main component of connective tissue. With the exception of brain tissue, fibroblastic cell types can be found throughout the human body. **Tenoblasts** are specialized fibroblasts found in the tendons [1].

Adipocytes are predominantly found in adipose tissue (fat) which surrounds most organs and tissues in the human body (e.g; Liver, Bone marrow etc).

2.1.4 Cellular Composition of Soft Tissues

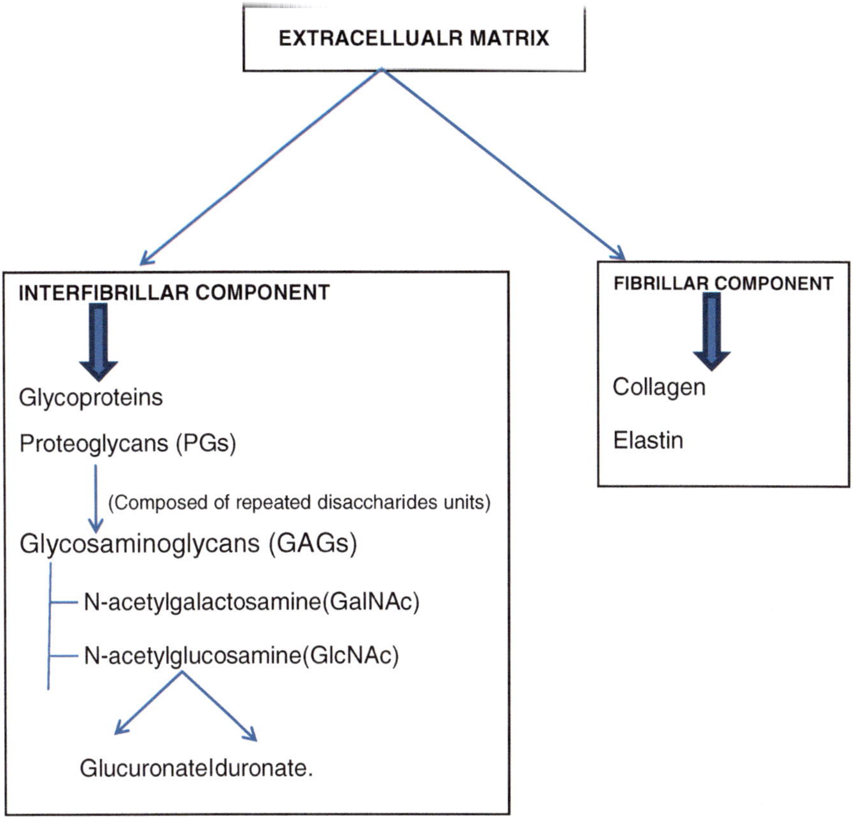

The composition of connective tissues determines its structure and functions which in turn relate to the biomechanics and physical properties. We shall learn about the behavioral properties of connective tissue in the upcoming chapter. Here, we shall focus on structural composition, and we find that the connective tissues are majorly composed of cellular component and the extracellular matrix. In muscles and nerves, the behavior of cellular component determines the biomechanical properties predominantly. The important cell types found in the tissues have been listed in Table 2.1 [1].

Among other connective tissues, the extracellular matrix determines the structure and functions. The extracellular matrix has interfibrillar and the fibrillar component. The interfibrillar component is composed of **glycoproteins** (protein with smaller chains of carbohydrates) and **proteoglycans** (protein with larger chains of disaccharide units known as the **glycosaminoglycans or GAGs**). The most important type of GAGs is listed in Table 2.2 [1].The fibrillar component contains two major classes known as the collagen and elastin.

Table 2.1 Cellular composition, location, and function in soft tissues

Cells types	Location	Function
Fibroblast	Found in tendon, ligament, skin, bone.	Creates mostly **type I collagen**
Chondroblast	Fibroblast found in cartilage	Produces mostly **type II collagen**
Osteoblast	Fibroblast found in bone	**Produces type I collagen and hydroxyapatite,** responsible for **bone regeneration**
Osteoclast	Found in bone	Responsible for **bone resorption**

Table 2.2 Common types and location of GAGs in the human body

GAGs	Location
Hyaluronan	Synovial fluid, vitreous humor, loose CT, healing CT, cartilage
Chondroitin sulfate	Cartilage, bone, heart valves, tendons, ligaments
Heparan sulfate	Basement membranes, cell surfaces
Heparin	Intracellular granules in mast cells lining arteries
Dermatan sulfate	Skin, blood vessels, tendons, ligaments
Keratan sulfate	Cornea, bone, cartilage

2.1.5 Composition of Important Connective Tissues

(a) **Ligaments**

Cellular component—10–20%
Extracellular matrix—80–90%
Interfibrillar component—dermatan sulfate <1%.

Fibrillar component—collagen type I predominantly, type III, type IV, and type V as well as elastic to a lesser extent (**ligamentum flavum** is an exception in having large elastic fibrillar components) [1].

(b) **Tendons**

Water—60–75%
Interfibrillar component—dermatan sulfate up to 1%.
Fibrillar component—80–95% type I collagen [1].

(c) **Bone**

Water—25%.
Interfibrillar component—chondroitin sulfate up to 50%
Fibrillar component—25–30% type I collagen [1].

The extracellular matrix of the bone is calcified and thus performs a specialized function. The inner layer is soft, known as the **cancellous or trabecular or spongy bone,** whereas the outer layer is hard and known as the **compact or the cortical**

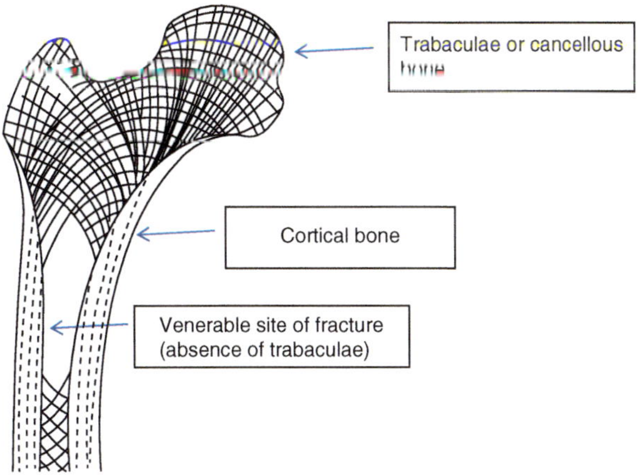

Fig. 2.1 Depicts the presence of cancellous and cortical bone in the femur

bone [1]. The cancellous bones under the stress and loading pattern form thin plates known as the trabaculae which help in load distribution evenly along the shaft and articular surfaces [1]. The density of the trabaculae depends upon the stress and loading pattern over a bone. The cancellous bone is covered by concentric layers of thin but hard surface which is known as the cortical bones as shown in Figure 2.1.

Clinical Implication: As we have learnt that the trabecular bones are responsible for optimum load distribution and thus can sustain the stress better than cortical bones. Now consider Fig. 2.1 where the trabaculae are less dense or absent. This suggests that the chances of fracture at such sites would be higher because the tensile stress of the bone is not optimal due to lack of trabaculae. It is the same reason why fractured neck of femur is most common.

2.1.6 Classification of Joints

In the above section, we have learnt about the composition of various connective tissues. These tissues ultimately function to provide mobility and stability to various joints of the body. Therefore, it is very important to know the different types of joints and classify them for easy understanding. The term *arthroses or articulation* has been used to explain the joint types based on the method and material of joining parts. Therefore, two major categories of joint classification include **synarthroses** (*non-synovial*) and **diarthroses** (*synovial)* [1].

2.1.6.1 Synarthroses

Joint formed using interosseous connective tissues like fibrous joints and cartilaginous joints.

1. **Fibrous joints: joints attached by fibrous tissue**
   (a) Suture—bones attached using collagenous suture ligament or membrane (e.g., skull)

   (b) Gomphoses—like a peg in a hole (e.g., mandible or maxilla)
   (c) Syndesmoses—bones attached using interosseous ligament, a fibrous cord, or an aponeurotic membrane (e.g., between tibia and fibula)
2. **Cartilaginous joints:** joints attached by fibrocartilage or hyaline cartilage
   (a) Symphysis joint—fibrocartilage forms disks or pads (e.g., intervertebral joints)
   (b) Synchondroses—attached with hyaline cartilage (first chondrosternal joint)

2.1.6.2 Diarthroses

Bones have distinct features such as a joint capsule that is composed of two layers, a joint cavity that is enclosed by the joint capsule, synovial tissue that lines the inner surface of the capsule, synovial fluid that forms a film over the joint surfaces, and hyaline cartilage that covers the surfaces of the enclosed contiguous bones [2].

2.1.7 Applied Biomechanics on Connective Tissues

We have seen that the composition of connective tissues determines their structure and functions; let us now understand how these apply to the biomechanics in context to connective tissues. It has been proposed that GAGs present in the connective tissues affect the hydration of that particular substance whether it be a bone tissue, muscle, tendon, or ligament [3]. The GAGs are hydrophilic in nature due to their negative charge which creates an osmotic pressure gradient for water inflow into the extracellular matrix. The water flow swells the interfibrillar component and creates a tensile stress on the surrounding collagen network where the collagen in reaction resists the swelling pressure, thereby maintaining the rigidity and stability of cells. The proteoglycans also work for nutrient supply, growth enhancement, providing strength to collagen, and maintaining the size of the collagen fibrils. The composition and content of GAGs also depends upon the type of loading on the tissues such as tensile loading or compressive loading. This could be the reason that tissues responsible for more compressive loads contain more chondroitin sulfate, compared to tissues taking more tensile loads which contain dermatan sulfate in abundance [4]. The glycoproteins help to fasten the extracellular matrix apart from adhesion of collagen and inner molecules of the cell membrane (e.g., fibronectin, laminin, tenascin, and osteonectin).

Among the fibrillar components, collagen is the most abundant protein comprising 25–35% of human body proteins [5]. The main function of collagen is to provide tensile strength for functional integrity of the connective tissues against the tensile forces [6]. Few important types of collagen such as types I, II, III, V, and XI are the most common. Type I is predominantly found in major connective tissues including the tendons, ligaments, bone, and synovium [7]. Type I collagen is mostly responsible for load-bearing properties subjected to tensile forces in particular, whereas the type II collagen fibers function to withstand compressive forces and thus are found abundantly in hyaline cartilage and nucleus pulposus of the intervertebral disk [8]. Type III collagen fibers are seen in the skin and synovium [9]. Type

V collagen is also found in the cartilage and tendon along with type XI which determines a significant kinetics and kinematics of the connective tissues.

Apart from the collagen, the elastin fibers contribute to significant load-bearing kinetic properties to the extracellular matrix. It has been found that ligamentum nuchae (one of the ligament of spine) comprises 75% elastic and 15% collagen, and thus, it has stretchable load sustaining capabilities.

Let us now understand how does the connective tissue behave and respond to the stress imposed on them. These concepts can help us to understand the abilities and limitations of the connective tissues in human biomechanics which could also be applied to our clinical practice.

2.1.8 Connective Tissue Response to Loading and Stress

The connective tissues respond to the imposed load and stress through deforming and reforming. The extent of change depends on the volume, nature, and frequency of imposed stress. It has been proposed that low-frequency compressive loading helps increase cartilage formation (slow mobilization of the knee and shoulder is key to cartilage healing in rehabilitation). On the other hand, higher frequency with compressive loading would enhance bone formation (in clinical practice after fracture, walking is important to load bone for callus formation). Alike, high magnitude of stress and sustained loading would cause fibrocartilage formation, whereas tensile loading would increase the strength of muscles and tendon (in clinical practice, we enhance muscle strength through optimal tensile loading). It should be noted that optimal and threshold point of loading is required for enhancing the performance through deformation and maintenance. Below the threshold deformation occurs negatively (as seen for ill effects of immobilization).

(a) ***Bone Response to Loading***

The bone response to loading is governed by **Wolff's law** which confirms that bone deposition increases with weight bearing. Osteocytes are the main cells that are responsible for bone development. In clinical practice, the best form of loading is exercise which is now very well adapted as the therapeutic intervention. It has been found that exercise can even maintain or regulate higher bone density in postmenopausal women [10]. In addition, the mechanical loading affects significantly all age groups where the growing skeleton in childhood and adolescence has maximum benefits. In adulthood loading can help to maintain the bone mass and density. Nevertheless, loading at aging and menopause can help minimize the bone loss and prevent osteoporosis [11]

(b) ***Tendon Response to Loading***

It is very evident that tendons respond to loading positively. At the composition level, the concentration of collagen increases with tensile loads which enhances the strength and stiffness. Chronic loading leads to hypertrophy as seen in muscles when trained with higher loads and repetition. In clinical practice, loading the tendon and muscles is commonly practiced to treat various musculoskeletal conditions. Even to increase the performance level of athletes, resistive training with loads is highly recommended.

Few studies have suggested that an application of tensile load leads to an increase in type I collagen (e.g; Eccentric heel drops in Achilles tendinopathy). In an ideal situation, there is a mechanostat point defined as a level at which loads can induce a positive or negative response, influenced by chronic load [12].

(c) ***Ligament Response to Loading***

The ligaments response to loading is constructive though a slower process than bone and muscles has been reported. There have been studies to demonstrate the positive effects of activity on ligaments [13]. However, the exact magnitude for loading and time course for positive adaptation is not very clear [14].

2.2 Summary

The composition of soft tissues plays a very important role in determining their structure and functions. The soft tissues of the human body are mainly composed of cellular component and the extracellular matrix. It is the composition of soft tissues that will determine the kinetic abilities and their stability function. The location of GAGs in the extracellular matrix of the soft tissues provides them with characteristic biomechanical features.

References

1. Levangie PK, Norkin CC. Joint structure and function: a comprehensive analysis. Philadelphia, PA: FA Davis Co.; 2011.
2. Allan DA. Structure and physiology of joints and their relationship to repetitive strain injuries. Clin Orthop Relat Res. 1998;351:32–8.
3. Heinegard D. Structure and metabolism of proteoglycans. In: *Extracellular matrix biochemistry*. New York: Elsevier; 1984. p. 277–328.
4. Yoon JH, Brooks R, Kim YH, Terada M, Halper J. Proteoglycans in chicken gastrocnemius tendons change with exercise. Arch Biochem Biophys. 2003;412(2):279–86.
5. Eyre DR. Collagen: molecular diversity in the body's protein scaffold. Science. 1980;207(4437):1315–22.
6. Vogel KG, Thonar EJM. Keratan sulfate is a component of proteoglycans in the compressed region of adult bovine flexor tendon. J Orthop Res. 1988;6(3):434–42.
7. Eriksen HA, Pajala A, Leppilahti J, Risteli J. Increased content of type III collagen at the rupture site of human Achilles tendon. J Orthop Res. 2002;20(6):1352–7.
8. Burgeson RE, Nimni ME. Collagen types. Molecular structure and tissue distribution. Clin Orthop Relat Res. 1992;282:250–72.
9. Liu SH, Yang RS, Al-Shaikh R, Lane JM. Collagen in tendon, ligament, and bone healing. A current review. Clin Orthop Relat Res. 1995;318:265–78.

10. Wolff I, van Croonenborg JJ, Kemper HC, et al. The effect of exercise training programs on bone mass: a meta-analysis of published controlled trialsin pre- and postmenopausal women. Osteoporos Int. 1999;9:1–12.
11. Bergmann P, Body JJ, Boonen S, Boutsen Y, Devogelaer JP, Goemaere S, Kaufman J, Reginster JY, Rozenberg S. Loading and skeletal development and maintenance. J Osteoporosis. 2011;2011:786752.
12. Lavagnino M, Arnoczky SP. In vitro alterations in cytoskeletal tensional homeostasis control gene expression in tendon cells. J Orthop Res. 2005;23(5):1211–8.
13. Gomez MA, Woo SLY, Amiel D, Harwood F, Kitabayashi L, Matyas JR. The effects of increased tension on healing medial collateral ligaments. Am J Sports Med. 1991;19(4):347–54.
14. Culav EM, Clark CH, Merrilees MJ. Connective tissues: matrix composition and its relevance to physical therapy. Phys Ther. 1999;79(3):308–19.

Application of Physical Laws on Human Body

3

3.1 Background

In this chapter, we will focus on the laws of physics applied to kinetics and kinematics of the human body. This would give us an understanding of how the mechanics of the human body is governed by the universal laws along with the few assumptions made.

3.1.1 Application of Force Vectors on Human Body

The definition and concept of vectors has been explained in Chap. 1. Here, we shall learn how to apply vector diagrams for human biomechanics. In the human body, for any joint segment, two vector forces act, i.e., horizontal and vertical, as shown in Fig. 3.1.

The horizontal component is always parallel to the ground, whereas the vertical component is perpendicular to the ground. The resultant as explained previously would be in the direction of the diagonal of parallelogram. Since this resultant would also have magnitude, you can use the Pythagoras theorem or laws of trigonometry to determine it. We shall learn these calculations in the kinesiology section. It is also important to know the frame of reference while considering these vectors, and any of the below could be used:

1. The vertical and horizontal direction relative to the ground
2. The planes of the human body, e.g., sagittal, coronal, or transverse
3. Along a body segment and at 90 degrees to it

The frame of reference relative to the ground is most suitable and applicable. However, if we use the other two also, the results would be similar. Considering the anatomical position of the human body, the frame of reference can be taken along

A. Hazari et al., *Conceptual Biomechanics and Kinesiology*,
https://doi.org/10.1007/978-981-16-4991-2_3

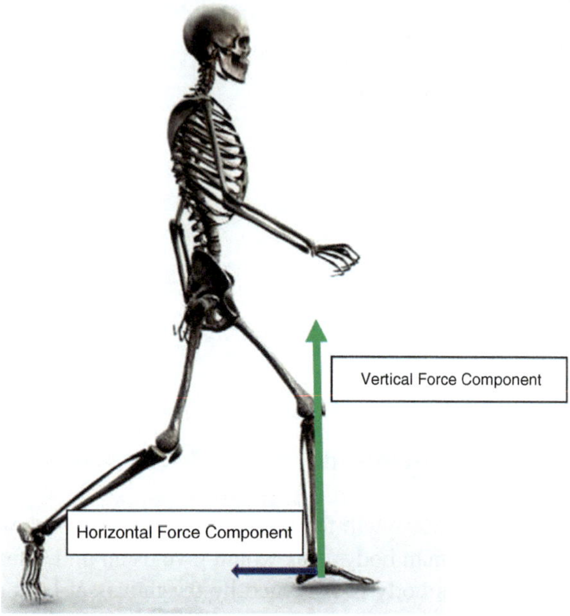

Fig. 3.1 Depicts horizontal and vertical component of force vectors

the long axis of bone and perpendicular to it. Let us see the vectors along the body segment with the help of Fig. 3.2.

If we look closely, the frame of reference taken in relation to ground and bony segment yields similar force vectors. Thus, in clinical practice, we can use them very frequently. The frame of reference used in relation to axis and planes are mostly done using the advanced motion analysis systems. In this book, we shall take bone segment or ground as the frame of reference wherever applicable.

3.1.2 Application of Newton's Laws of Motion

The Newton's laws of motion are universally applicable in all physical bodies. They help us to identify kinetics and kinematics with accuracy.

3.1.2.1 Newton's First Law

The first law, also known as the law of inertia, states that a body at rest would remain at rest and a moving body would keep moving until an external force acts to change its state of rest or movement (inertia). However, this law is not totally applicable to the human body as we constantly experience some external force or resistance to movement. For example, if you are walking, you experience the frictional force which disturbs the state of constant dynamics, suggesting that if there was no friction you would not have stopped and kept moving. In human beings, if the sum of the forces acting on the body is zero, the body would be at rest. In order to break this state of inertia, you either need a push/pull from an external environment or

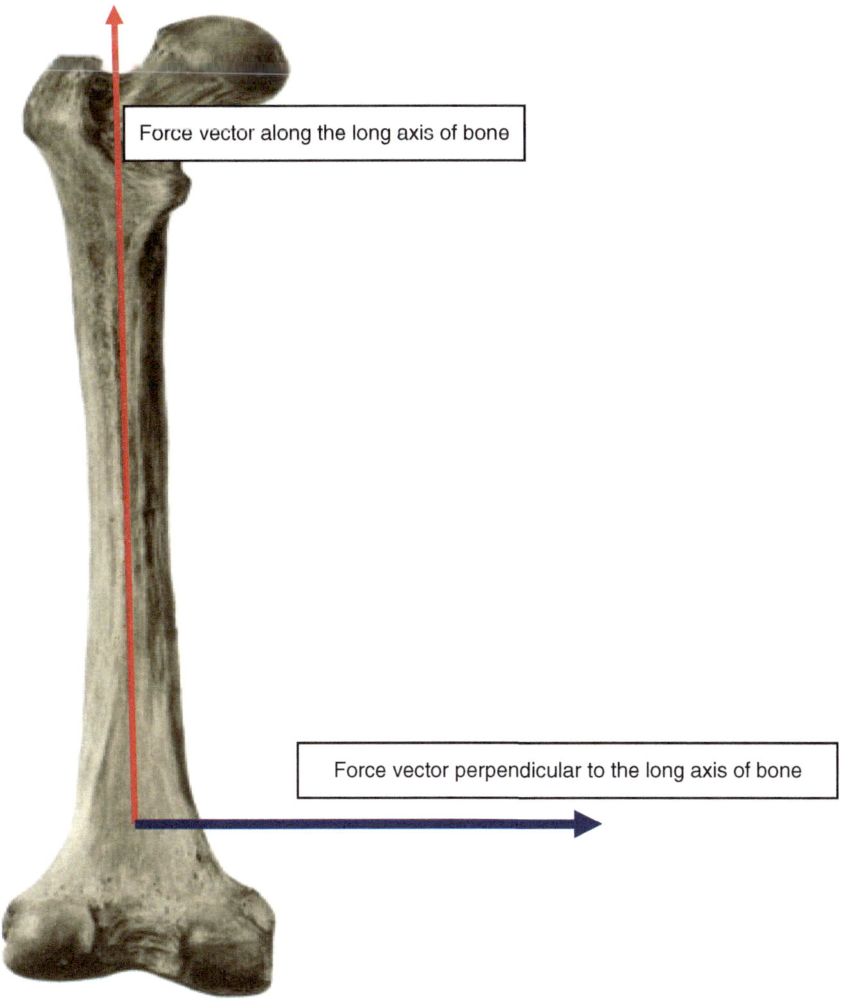

Fig. 3.2 Depicts the force vector in reference to the body segment (femur bone—along the long axis and perpendicular to the long axis)

muscular internal force. Thus, as per the first law of Newton, the human body is guided by the state of inertia, but internal and external forces maintain the state of statics and dynamics.

3.1.2.2 Newton's Second Law

In the above section, we learnt that there is a force that maintains the state of inertia. The magnitude and direction of this force is governed by the Newton's second law. The law states that the force acting on the body is the product of its mass and acceleration in the direction of the movement (F = mass × acceleration). For example if you are pushing a box, the force produced is the product of mass of the box and rate of change of its velocity (acceleration). The newton's second law can be applied on human body segments to calculate the muscular forces. In biomechanics, multiple forces act on a single point, and in order to determine the magnitude and direction of motion, we need to resolve all force into a single force vector.

3.1.2.3 Newton's Third Law

The third law governs the principle of equal and opposite reaction to a force applied. This means, if you apply an internal force on the human body, there would be an opposite and equal external force. The ground reaction force and joint reaction force are the most suitable applications of the third law in human biomechanics. They form the basis for human kinetics and kinematics.

3.1.3 Application of Load on Connective Tissues

The connective tissues have elongation properties. Any load applied to connective tissues causes a **force–elongation** relationship which we shall discuss here [1]. In context to the tensile loading, the connective tissue creates different responses according to the magnitude of loading and its duration which is known as the **visco-elastic** properties. Before we focus on viscoelasticity, let us understand the normal force–elongation curve as depicted in Fig. 3.3.

The load applied on the connective tissues helps us determine the strength of the tissues, its elongation capacity **(elasticity),** permanent deformation **(plasticity),** and the **failure point**. Let us understand this with the illustration above which depicts a graphical representation between the load and elongation. We can see that the graph has a distinct **bell-shaped** curve which suggests that the load causing an initial increase in elongation flattens at a certain point followed by a decrease due to tissue failure. The area represented under the region A is known as the **toe region,** where a load applied leads to a proportional increase in the connective tissue lengthening

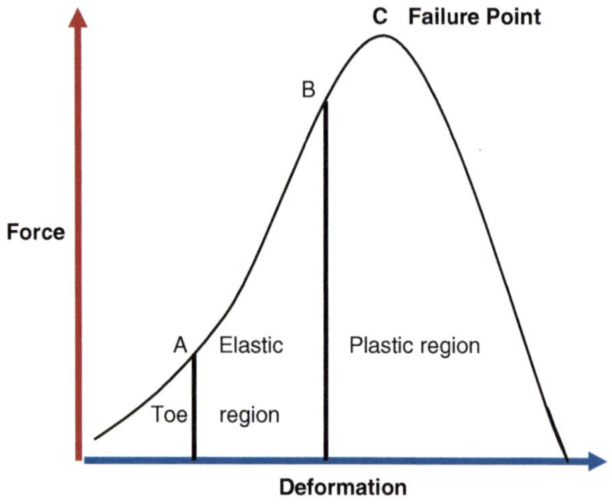

Fig. 3.3 Depicts force–deformation curve for a connective tissue

through the straight line/slope observed (please note that the elongation here is seen at the microscopic level in the contractile elements of the connective tissue).

The region covered under points A and B is known as the **elastic region** which determines that the changes in tissue deformation will not be permanent and the connective tissue would return to its original shape and size after removal of the applied load. There is an exponential increase in length with load in this region, and it determines the stretchability of the connective tissue. Clearly, the more the elastin component, the better would be elasticity and more area would be covered by the connective tissue in this region. Therefore, tendon and muscle have the ability to stretch more compared to bone in their respective elastic regions. The end of the elastic region has point B, known as the **yield point**, beyond which the connective tissue loses its elasticity. The area represented within points B and C is known as the **plastic region**. The connective tissue in this region undergoes permanent change while a load is applied on it and thus does not return to its original shape and size although the material does not give up. Load beyond this point leads to failure of the tissue at point C, known as the **failure point**.

The deformation seen in the connective tissue on application of load is governed by the laws of Newton. The change or deformation is consistent with Newton's law of inertia. The magnitude of force that determines the extent of deformation **(strain)** in terms of size, length, composition, etc., is governed by the second law, whereas internal response to deforming force is known as the **stress** calculated as force in Newton (N)/ per unit area in meter square (N/m^2) [2]. For example, if a force of 50 N is applied on a tissue of cross-sectional area 5 m^2, the stress would be **10 N/m^2** (50 N/10 m^2).

The strain is the relative change or deformation seen in the connective tissues and calculated as the difference in the original and deformed length in relation to the original length. For example, consider that the original length of a tissue is 10 meter and a load of 50 N is applied on it. The new length of the tissue is 20 meter due to the elongation. Therefore, the strain would be 20 m (new length)—10 m (original length)/ 10 m (original length) = **1%**. It can be interpreted that a 50 N force on a 10 m tissue can lead to 1 percent strain.

The stress and strain developed in the connective tissue would depend upon the following:

(a) The material composition (dealt with in Chap. 2)
(b) The type of load (compressive loading, tensile loading, shear force will cause compressive, tensile, and shear stress and strain, respectively. Description about these forces can be seen in Chap. 1).
(c) Point of load application, direction, and magnitude as governed by the Newton's law
(d) The rate and duration of load as explained by the load deformation curve (viscoelasticity)

When we consider the factors altering the stress and strain of connective tissue, we find that the material internal resistance also known as the **stiffness** holds a significant value. The stiffness of the material is represented by a physical quantity known as the **Young's Modulus.** In a stress–strain curve, we find that the Young's modulus can be determined as the slope of the curve or tangent.

It is evident that the steeper the slope, the higher would be the Young's modulus and material stiffness. Thus, the ability of such tissue to bear the load would be higher due to its stiffness. For example, the cortical bone is stiff because of its higher modulus of elasticity (Young's modulus). In contrast, a gradual slope would suggest that the material is less stiff which is also known as the compliance of the material. It should be noted that the stiffness of connective tissue is an internal property which does not depend on the size of the material.

Let us try to understand the stress–strain curve of muscle injury. We already know that the muscle injuries (strain) have grading where Grade I strain consists of injury to few muscle fibers, Grade II sprain is seen as partial tear, and Grade III is seen as complete rupture. On the load deformation curve we would find that Grade I injury could be seen in the elastic range, Grade II in the plastic range, and Grade III at the failure point. The stress–strain curve can be seen to become steeper when we move from elastic to plastic range, suggesting the higher stiffness of the material and resistance to change which ultimately would give away at the failure point.

3.1.4 Application of Viscoelastic Properties of Soft Tissue on Human Biomechanics

We have seen that connective tissue responds to load by deforming, but the change is rate and time dependent which determines the **viscoelasticity** of the material. As the name suggests, the viscoelasticity of any material is a combination of viscosity and elasticity. The viscosity is the resistance of the material to deform, and elasticity is the ability of the material to stretch. As per the physical law for conservation of energy, the energy produced during deformation due to elastic properties of the material will be used back to return to its original state due to its viscosity although some amount of energy would be dissipated.

Considering the connective tissues, let us understand the viscoelasticity applied to muscle. The muscle contraction has been classified majorly into **concentric** and **eccentric.** The eccentric contraction where the muscle stretches represents the elastic property. The energy produced as the work is done against the muscle viscosity is returned back when the muscles contract in concentric action. Also, it has been found that the viscoelastic properties of connective tissue are time and rate dependent [3].

3.1.4.1 Time-Dependent Viscoelastic Properties

We have learnt that any tensile force or compressive force would cause deformation in the connective tissue depending upon its elastic abilities, but the viscosity of the material makes the deformation time dependent. Let us learn some time-dependent variables of viscoelasticity.

(a) **Creep:** The connective tissue responds to a time-dependent phenomenon called as creep. When we apply a constant load/force on the connective tissue, the deformation would increase with increased time which represents the creep.

Clinical Application 1: Stretching of Muscles
While doing a muscle stretch, we apply a constant tensile force and hold it for a specific period of time. We find that the muscle has lengthened over time (creep) with that constant loading. But when we remove the load the muscles again retain its original state which is known as the recovery.

Clinical Application 2: Synovial diffusion
When we apply the compressive force on cartilage and ligament, the structures deform and there is synovial diffusion for nourishment of connective tissues.

(b) **Stress-Relaxation**

In contrast to creep where load is constant, if the tissue is stretched and length is kept constant, the load or force required to maintain the stretch decreases over time which is known as the stress-relaxation. As the name suggests, the tissue is stressed initially which relaxes over time.

Clinical Application—When we stretch a muscle and hold the length constant, we find that the force applied to initially stretch the muscle to that specific length reduces as the tissue relaxes.

3.1.4.2 Rate Dependent Viscoelastic Property

(a) **Strain-Rate Sensitivity**

The strain-rate sensitivity depends on the rate at which the load is applied to the connective tissues. The connective tissues can sustain higher loads if applied rapidly in comparison to load applied slowly. Consequently, the stress relaxation would be higher and creep would take longer when the load is applied rapidly.

Clinical Application: when you need to transfer a heavy load from one place to another, you apply the force in shortest time in lifting and transfer.

(b) **Hysteresis**—Hysteresis is a physical term coined for the loss of energy. In terms of human biomechanics, the load deformation curved produced by a load does not lead to full return of energy when load is removed and some amount of energy is lost as heat in the recovery process. This process is called as hysteresis.

Clinical Application: If we stretch a muscle eccentrically the muscle does not recover with the same energy concentrically and some energy is lost as heat.

3.1.5 Viscoelastic Response of Soft Tissues on Loading

We have learnt the viscoelastic properties of soft tissues, and now we shall focus how these properties are exhibited when the soft tissues like bone, muscles, ligament, and cartilage are loaded. The stress–strain curve for each soft tissue is significantly different which suggests that the load-bearing capacity and ability to sustain deformation would be different for each soft tissue. The understanding of such concepts can help us determine the extent to which we should be using the applied force in therapeutic interventions based on the soft tissue that is being targeted.

3.1.5.1 Viscoelastic Response of Bones

The cortical bones have higher modulus of elasticity and thus stiffer than the cancellous bone suggesting that cortical bones can take more stress/load, whereas cancellous bone can take more strain. In vivo experiment has demonstrated that cancellous bone can withstand up to 75% strain compared to cortical bones which can only sustain 2% before failure [1]. The cortical bones have higher compressive load and tensile load sustaining capabilities in the longitudinal axis than in the transverse axis. However, the compressive loading is better taken than tensile loading in cortical bones. This is the reason that transverse fracture of the femur is more common even with a smaller magnitude of force perpendicular to the long axis. The cancellous bones can also sustain more stress with lesser strain in compression compared to tensile loading [4]. As we have learnt, the bones respond positively to applied load with osteoblastic activity, the compressive loading on cancellous bones leads to an increase in the number of cells and thereby causes hypertrophy. In regard to the viscoelastic property, bones lose their stiffness and strength when loaded repetitively with high frequency or high load due to strain when the bone is creeping.

3.1.5.2 Viscoelastic Response of Muscles and Tendons

Creep is very commonly seen in tendons when muscle is subjected to a constant loading. The muscles contract without change in length (isometric) or in a lengthened position (eccentric). Once the length is increased through creep, less force is required to maintain the original length (stress-relaxation). In the toe region of the load–deformation curve, the muscle undergoes 2% strain, and increased load in the elastic region would show a linear increase in the strain where the continued load could damage few muscle fibers in the plastic range. Ultimate failure and rupture of muscle would take place for further loading beyond the failure point. The cross-sectional area of the muscles plays an important role in the stress and strain bearing capacity of muscle, and thus, longer muscles or tendons can withstand higher loads. However, the composition of soft tissues should also be taken into consideration. On optimal loading, the muscles thickness (hypertrophy) and strength increases. On the other hand, the strength and thickness reduces (atrophy) when loading is less than optimal such as seen during immobilization.

The muscles and tendons are a continuous unit; however, the tendons are subjected to more stress because of the composition, especially at the insertion to the bone, but they can withstand considerable tensile loading [1]. In clinical practice, muscles and tendons exhibit creep, stress-relaxation, and strain rate sensitivity which makes them functionally better with rehabilitation and progressive loading.

3.1.5.3 Viscoelastic Response of Ligaments

Similar to tendons, ligaments respond well to the tensile loads. In addition, ligaments have the ability to withstand stress in multiple directions [1]. The ligament strength increases with optimal loading, and under loading has a negative effect. In contrast to tendons, ligaments can take compressive, tensile, and shear force more efficiently.

3.1.5.4 Viscoelastic Response of Cartilage

The cartilage can resist the applied load by developing stress in the fibrillar component, swelling pressure, and frictional drag [1]. The compressive force reduces volume of the cartilage as the interstitial fluid moves out leading to deformation until the applied load is well balanced by the resistance created by the cartilage. The tensile stress produced in cartilage is similar to that seen in the muscle and ligament; however, a nonlinear graph is seen in the toe region. The shear stress at cartilage depends upon the collagen concentration.

3.2 Summary

The physical laws and principles governing the motion and forces on human body are useful to understand the kinematic and kinetic behavior. The universal Newton's laws are also applicable to human biomechanics as the basis for understanding the different forces acting on a human body. It is the applied physical laws which keep the body stable through an application of internal and external force. The soft tissues of human body respond differently to the load applied depending upon their composition.

References

1. Levangie PK, Norkin CC. Joint structure and function: a comprehensive analysis. Philadelphia, PA: F.A. Davis; 2011.
2. Goldstein SA, Armstrong TJ, Chaffin DB, Matthews LS. Analysis of cumulative strain in tendons and tendon sheaths. J Biomech. 1987;20(1):1–6.
3. Özkaya N. In: Özkaya N, Nordin M, Leger DL, editors. Fundamentals of biomechanics: equilibrium, motion, and deformation. Cham: Springer; 1999.
4. Nordin M, Frankel VH, editors. Basic biomechanics of the musculoskeletal system. Philadelphia, PA: Lippincott Williams & Wilkins; 2001.

Applied Biomechanics on Joint, Muscle, Tendon, and Ligament

4

4.1 Background

In the previous chapters, we have learnt the basic concepts of biomechanics which is applied to the human body through universal physical laws. In this chapter, we shall learn about the applied biomechanics on the individual soft tissues in context to their mechanics of movement and functioning. These concepts help to determine the physiological behavior of the soft tissues and correct the pathomechanics associated with it. We will focus on the kinematics and kinetics of the joint, muscles, tendon, ligaments, and capsule.

4.1.1 Applied Kinematics at Joint

In Chap. 1, we introduced arthrokinematics which is related to the motion of the joint. In human biomechanics, the joint motions are predominately considered to have two motions, viz. **(a) rolling** and **(b) gliding.** Rolling is considered as the **rotatory motion,** whereas gliding is represented with **translatory motion**. It is important to understand the frame of context for determining the direction of rolling and gliding. In order to ascertain the direction of **rolling**, we take the **distal segment of the bone** and for **gliding** we consider the **proximal segment** of the joint. For example, if we want to apply these concepts on the shoulder joint, we would consider rolling in the direction where the humeral condyle moves and gliding in context to the head of the humerus. However, it should be noted that both rolling and gliding are occurring at the shoulder joint (head of the humerus and glenoid cavity) and reference to distal and proximal segment is for understanding the concept. Given that we have now understood the arthrokinematic reference to direction, we will now learn about the **convex–concave rule** and vice versa.

Convex–Concave rule—The convex–concave rule suggests that when a convex surface is moving on the concave surface, the direction of **rolling and gliding is**

A. Hazari et al., *Conceptual Biomechanics and Kinesiology*,
https://doi.org/10.1007/978-981-16-4991-2_4

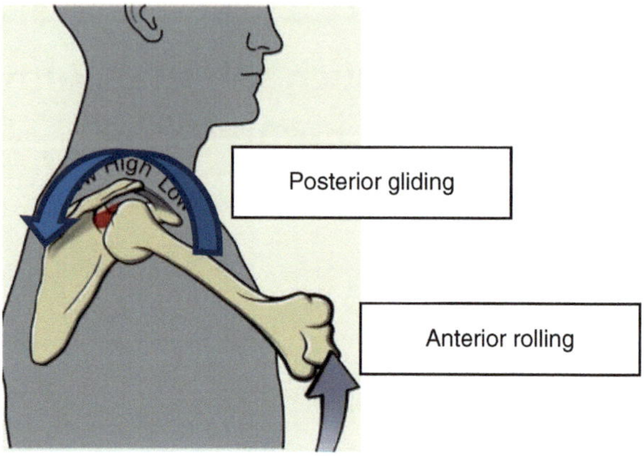

Fig. 4.1 Depicts rolling and gliding for shoulder flexion

apposite to each other. For example, when we do shoulder flexion, the direction of rolling is anterior, which can be taken in reference to humerus distal end and gliding would be posterior as seen in Fig. 4.1. The head of the humerus is the convex surface moving over the glenoid cavity which is the concave surface.

Concave–Convex rule—In contrast to the convex–concave rule, when the concave surface moves over the convex surface **rolling and gliding take place in the same direction**. For example, in high sitting position when the tibia which is the concave surface moves over the femoral condyles (convex surface) for extension, rolling and gliding both occur in the anterior/forward direction as shown in Fig. 4.2.

We have seen that the applied biomechanics on the joint depends upon the type of articular surface as discussed above. The other major factor that determines the arthrokinematics is open and closed kinematic chain movements. In the example given above for knee extension where the tibia is free to move over the femur in open kinematics, the direction of rolling and gliding is followed as per the concave–convex rule, but if the same motion occurs in close kinematic chain as seen during squats, the femur would move over the fixed tibia and the arthrokinematics would be governed by the convex–concave rule (Fig. 4.3). Thus, while determining the rolling and gliding movement directions, it is also important to check the reference of movements in context to open and closed kinematics.

4.2 Applied Kinetics at Joints

We have learnt that among all kinetic variables, the ground reaction force is the most considerable which gives rise to other types of joint forces. In accordance with the Newton's third law, the body weight creates an equal and opposite force arising from the ground. Due to the skeletal complexity and configuration, the ground reaction force is not equal at all joints and determined by the distance of line of gravity passing through the joint center which creates a torque or rotatory force on the joints. This is termed as external moment and also discussed in Chap. 1 as well as in kinesiology section. The magnitude of this torque is the product of force and

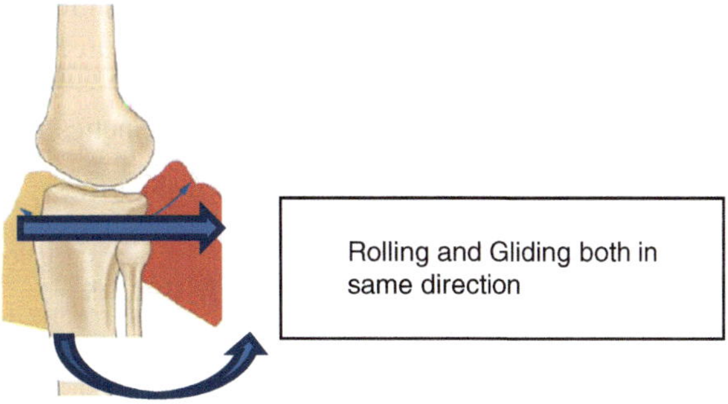

Fig. 4.2 Depicts directions for rolling and gliding as per the concave–convex rule at knee joint (open chain kinematics)

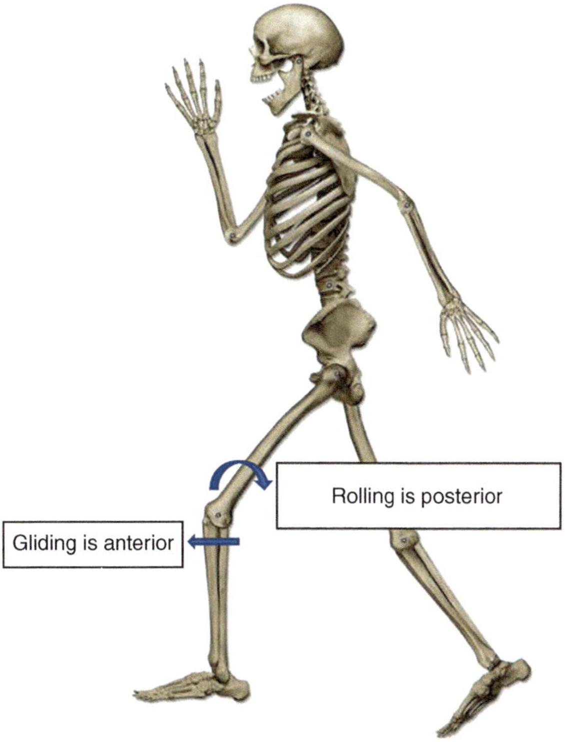

Fig. 4.3 Depicts directions for rolling and gliding as per the convex–concave rule at knee joint (closed chain kinematics)

perpendicular distance of the joint from line of gravity. This suggests that joint centers which are at larger distance from the line of gravity would experience larger torque compared to joint centers with smaller distance having similar mass.

4.2.1 Clinical Application of Forces on Joints

In clinical practice, we apply different forces on the joints for maintaining physiological functions. The compressive force at joints induces stability, osteoblastic activity, and supply of nutrients. The distraction force can help to mobilize the free radicals and clear waste products. The shear force can be effectively used for joint play and the frictional force at the joints helps to check hypermobility. Therefore, in clinical practice, the application of joints kinetics is of essential use to obtain the desired outcome.

4.3 Applied Kinematics on Muscles

We all know that muscle is a contractile soft tissue and their motion is determined by either shortening or lengthening. Muscles are attached to the bone and their action is categorized into three distinct types, viz. (a) **concentric contractions, (b) eccentric contraction, and (c) isometric contraction.** The physiology behind the muscle contraction can be referred to as **"actin-myosin filament theory."**

(A) **Concentric actions of muscle**—When the muscle is contracting while shortening, the action is known as the concentric contraction. The origin and insertion point of the muscles come close together. For example, the biceps brachii contracts concentrically during elbow flexion. Muscle contractions under concentric category mainly perform the action of **agonist or prime movers.** In other words, the muscles perform the actions to bring the joints in close proximity creating smaller acute angles as shown in Fig. 4.4.

(B) **Eccentric action of muscle-** When the muscle is contracting while the length of muscle is increased, the contraction is known as eccentric. During the eccentric contractions, the muscles increase the obtuse joint angle. These muscles mainly perform the activity as **antagonist** and joint stabilizers. An example of the eccentric contraction would be seen with the quadriceps contraction while squatting. When the knee joint flexion occurs, the length of the muscle increases though muscle is contracting to hold the squat position as shown in Fig. 4.5.

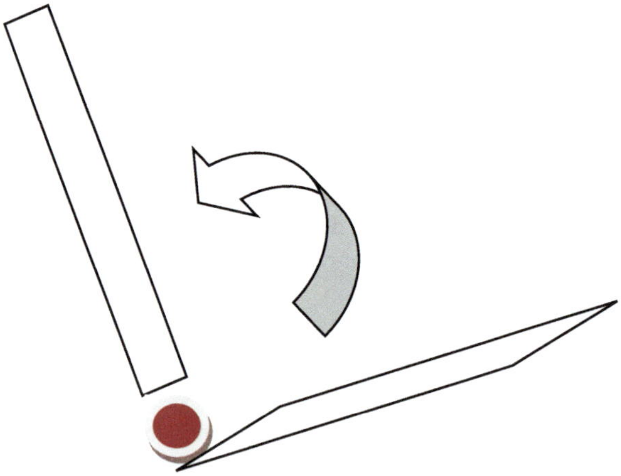

Fig. 4.4 Depicts concentric contraction of biceps at the elbow joint (the angle between the humerus and radius/ulna reduces)

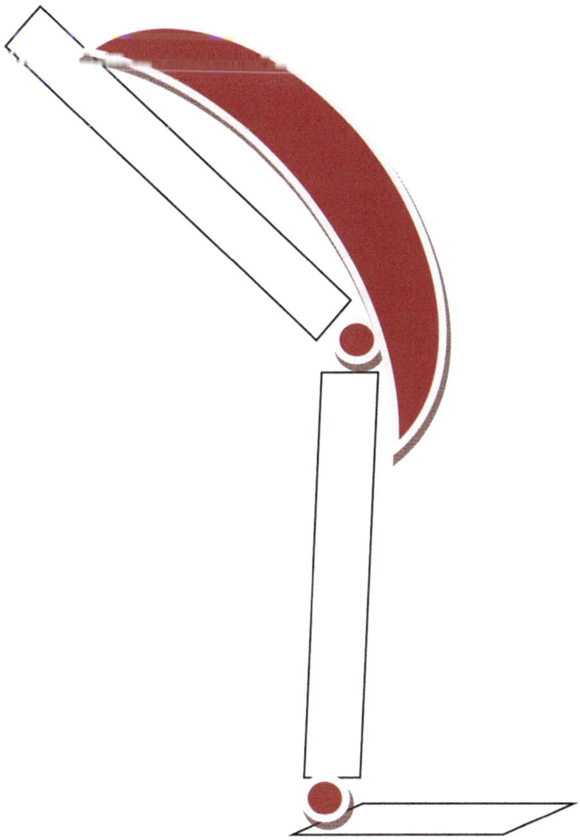

Fig. 4.5 Depicts eccentric contraction of quadriceps at the knee joint during squat

(C) **Isometric action of muscle:** The term isometric contraction has been given to define the muscle action while the length remains constant. Suppose you have been asked to lift a weight in both palms and keep the elbow bent constantly for 10 minutes at 90 degrees. The biceps muscle would contract concentrically from full extension to elbow flexion at 90 degree. However, when we hold the position for 10 minutes, the biceps keep contracting but the length of the muscles is constant and not changed. This is referred to as isometric contraction as shown in Fig. 4.6.

4.4 Applied Kinetics on Muscles

We understand that muscles primarily generate force to move joints and body segments. The magnitude of the force generated by muscle is governed by the neural input. For example, it is neurally driven that the force required for lifting a glass of water would be less compared to lifting a heavy weight. On the other hand, biomechanically the force generated by muscles is determinative. For instance, we have

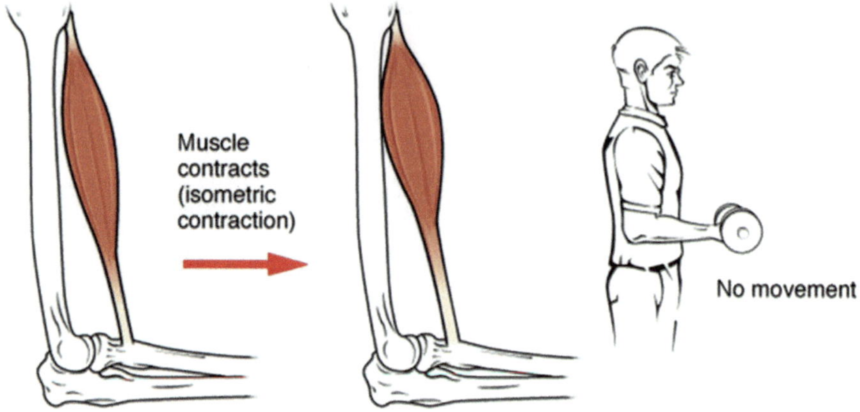

Fig. 4.6 Depicts isometric contraction of biceps at the elbow joint

learnt that the ground reaction force creates an external moment on the joints and in response the muscle needs to create an internal moment or torque to counter the force to keep the body in equilibrium. The force generated by the muscle would change depending upon the state of the body either in static or dynamic equilibrium. The magnitude and direction of the force generated would depend upon the following factors:

(A) Point of muscle attachment to bone
(B) Number of joints crossed by the muscle
(C) Length–tension relationship
(D) Types of muscles and muscle fibers
(E) Neural input and recruitment pattern

Let us discuss each in detail

(A) **Point of muscle attachment to bone**—The action of the muscle is directly related to its point of location and attachment to bone. The most convenient way to determine the action and direction of muscle force is to imagine the muscle contracting from its **point of insertion to origin** (prime movers or agonist for concentric action). The muscle will pull the bone from the insertion point toward its origin in a very similar manner like you pull a weight attached to a rope standing from a point where the rope is attached to its starting end. An illustration of biceps and triceps muscle action has been given in Fig. 4.7.

It is very evident that if we try to pull the biceps muscle from its insertion point at the forearm toward the origin, it would lead to elbow flexion and similar application for triceps would cause elbow extension. It should also be noted that the

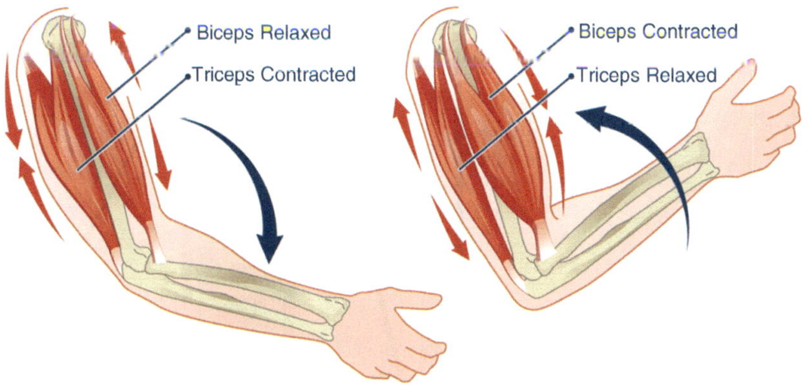

Fig. 4.7. Depicts muscle action of biceps and triceps

muscles attached to the anterior aspects of the joints are mainly flexor groups, whereas the muscles attached to the posterior aspects behave as the extensor group of muscles. Similarly the muscle attached medially would act as adductors and abductor when attached laterally. The contraction of one group would lead to relaxation of the other during concentric contraction (Fig. 4.7). However, during eccentric contraction both groups of muscle are acting in co-contraction to stabilize the joint. Now the direction of the muscle action and its magnitude is determined by the resultant force vector and angle to which it is attached to the bone. We shall learn these aspects and mathematical calculation of force in the kinesiology part of this book.

Clinical Application: In context to the magnitude of muscle force or strength for clinical application, it is very much important to maintain the strength ratio between the agonist and the antagonist. For instance, you have learnt about the Q/H ratio, i.e., quads to hams strength ratio, and when there is an injury to knee, the rehabilitation goals are to maintain the strength in desired ratio; otherwise, the chances of frequent injuries would be higher. It is also important to train muscle in the functional range and activity to train muscles in their normal physiological pattern.

(B) **Number of joints crossed by the muscle:** The muscle action and magnitude of force also depends upon the number of joints a muscle is crossing such as single joint or two joints. It has been proposed that human kinematics is quite complex, and in order to produce smoother movements, a coordinated muscle action is required where various muscles perform a single function to accomplish it efficiently through superior mechanics than robots. This is possible when the muscles attached to single joint could produce enough strength or power accompanied by two joint muscles that produce stability. Therefore, the single joint muscle is very efficient in performing concentric and isometric contractions, whereas the two joint muscles work to make the joint stable in dynamic motions [1].

Fig. 4.8 Showing length–tension relationship of muscles

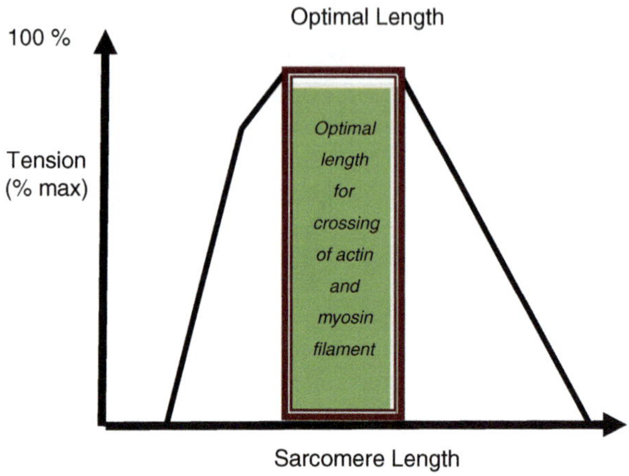

Clinical Application: It should be noted that the complex movement (multi-axial) in human biomechanics is a combination of one joint and two joint muscles. For example, the biceps is a two-joint muscle that performs supination and flexion at elbow but is well stabilized by the compressive force of the brachioradialis which is a single-joint muscle.

(C) **Length–tension relationship:** The length–tension relationship in muscles is one of the most important factors determining the magnitude of the force produced. The basis of length–tension relationship in the skeletal muscle was well established [2]. There are multiple models to explain the concept. Let us understand this with the graphical representation in Fig. 4.8.

The graph above represents the percent of maximum force (tension) against the length of the muscles represented by its sarcomere. Physiologically the tension produced in the muscles is dependent on sliding of its myosin and actin filaments. From the figure above, it can be clearly seen that the tension in the muscle is maximum at the optimal length of the muscle fiber, suggesting that if the length is short (lesser than 70% of its optimal length) or overstretched (more than 170% of optimal length) the force produced would be less. While the muscle is shortened, there is too much overlap between the actin and myosin components whereas under the overstretched state there is no overlap. Thus, in order to produce the peak tension an appropriate cross-bridging is required.

The length–tension relationship of muscle is also useful in determining the magnitude of tension at different stages of sarcomere length as it can be seen that the relationship is nonlinear. With an increase in muscle length (stretch), the tension increases until it reaches a plateau beyond which the tension decreases while length increases which can be seen as an inverted U on the graphical representation. For quantitative approach, the length–tension relationship can be expressed as the equation given by Hooke's law ($\Delta F = k.\ \Delta l$). The change in muscle force (F) depends upon the change in length (l) and its constant (k).

Clinical Application: The muscles like biceps generate maximum force in the mid range compared to shorter or longer range of motion.

The magnitude of force does not only depend upon the length of muscle; rather, the velocity of muscle also has a role to play represented by the force–velocity relation similar to length–tension. In a faster movement, the time required for muscles to overlap decreases; thus, less force is generated, whereas if the velocity is slow for adequate cross-bridging, higher force can be taken by the muscles. This is the basis for muscle strengthening techniques against the resistive loads; since higher forces produce greater strengthening, the velocity of muscle should be slow and steady.

(D) **Types of muscles and muscle fibers:** The human body is composed of various muscle types such as skeletal muscles, cardiac muscle, and smooth muscles. In context to biomechanics, we shall concentrate here on the skeletal muscle types and its fibers. The magnitude of force generated at joint depends upon the classification of skeletal muscles. In this chapter, we shall learn the chemical, metabolic, and physiology basis of skeletal muscle classification.

Under the metabolic classification, the ATPase activity is used to identify the muscle fiber types. The major fibers include **type I (slow fibers), type II A (intermediate), and type II B (fast fibers).** The metabolic basis identifies three main types of muscles such as **fast-twitch glycolytic, fast-twitch oxidative glycolytic, and slow-twitch oxidative**. It should be noted that human muscles contain both oxidative and glycolytic fibers and the muscles are classified based on the higher proportion of their oxidative or glycolytic composition. The oxidative function allows the muscle to work longer with smaller force and sustain the action without getting fatigued easily as seen in the type I fibers. On the other hand, the glycolytic components represent burst of higher force but shorter sustainability (Type II B), and the oxidative glycolytic muscles have equal composition. For example, the gastrocnemius muscles have 50% of oxidative and glycolytic fibers each thus can generate significant force for a longer period of time [3]. In contrast, the soleus and diaphragm muscles have more of oxidative components and thus classified as type I for their longer action without getting fatigued. The oxidative muscles thus function as stabilizers and are also known as postural muscles. Glycolytic muscles like hamstring (type II) that generate high force are also known as mobility or non-postural muscles.

Clinical Application: High-intensity training targets glycolytic or type II muscle groups, whereas low-intensity training should be used to train oxidative muscles.

Based on the physiological classification, the skeletal muscle is identified by their size and arrangement. The size of the muscles by its length determines the contractibility and resultant force produced. A muscle which is longer in length would produce more force compared to shorter muscle if the origin is same and attached to bone at the same angle as the mass of the longer muscle would be more. This is the reason in ideal conditions that a taller fast bowler can produce greater force through his shoulder muscle compared to a short stature fast bowler. The technique also plays a role; thus, the concept should be taken under the experimental conditions.

The other factor that determines the muscular force generation is its cross-sectional area and their arrangement. Muscles with more cross-sectional area would have more number of myofibers and motor units; thus, force generation would be high. For instance, the quads have higher cross-sectional area compared to hams; thus, quads are normally stronger group of muscles producing higher force. If the muscle fibers are arranged parallel to each other and the long axis of bone, they are known as **strap or fusiform muscles** (sternocleidomastoid, rectus abdominis, quads, and hams) [3]. Muscle fibers having the ability to rotate and twist the muscle fibers are **spiral muscles** (supinator). If the muscle fibers are arranged oblique to the long axis of bone and each other, they are called as **pennate** muscles. The common example for unipennate includes flexor pollicis longus. Muscles such as gastrocnemius and subscapularis are bippenate and multipennate, respectively.

(E) **Neural input and recruitment pattern:** A functional neural bimodal signals are the utmost necessity for stronger and smoother connection between the nervous and musculoskeletal systems. The neural inputs and recruitment pattern of muscles are major factors for determining the strength and quality of force produced at the joint. In pathology like stroke, the muscles lose their strength due to lack of established neural pathways.

4.5 Summary

In this chapter, we have learnt the basis for application of biomechanics to soft tissues which determine their function. We are now ready to apply these concepts to the peripheral joints to learn their biomechanics and kinesiology in detail.

References

1. van Ingen Schenau GJ, Dorssers WM, Welter TG, Beelen A, De Groot G, Jacobs R. The control of mono-articular muscles in multijoint leg extensions in man. J Physiol. 1995;484(1):247–54.
2. Gordon AM, Huxley AF, Julian FJ. The variation in isometric tension with sarcomere length in vertebrate muscle fibres. J Physiol. 1966;184(1):170–92.
3. Levangie PK, Norkin CC. Joint structure and function: a comprehensive analysis. Philadelphia, PA: F.A. Davis; 2011.

Principles of Kinetics and Kinematics on Human Body

5

5.1 Introduction

The human biomechanics is a complex subject and governed by certain principles pertaining to kinematics and kinetics apart from the physical laws that we have learnt in the previous chapters. It is the ability of the human body to dissociate one segment with other and apply the principles of biomechanics over the target body segment. For example, complete shoulder kinematics and kinetics is required to deliver a fast ball, while only the wrist and elbow need to be targeted for playing dart. Thus, the body mechanics strictly follows the principles of kinematics and kinetics which we shall deal with in this chapter.

5.2 Range of Motion Principle

The principle for range of motion states that less range of motion is most effective for low-effort (force and speed) and high accuracy movements, while greater range of motion favors maximum efforts related to speed and overall force production [1, 2]. This can be explained with an example discussed here. A carom player limits his range of motion at wrist and fingers to accurately and precisely produce the force required to hit the coins and move in specific direction. This suggests that less range of motion is required for more accuracy with low efforts. On the other hand, a baseball player uses a larger range of motion starting from the lower limb and transfers to the shoulder for a high velocity throw. Similar principle is used by cricket fast bowlers where they run and gain range of motion to be applied to the shoulder through a series of muscle actions. This is known as the transfer of momentum. Biomechanics studies help us to apply the principles of range of motion through the qualitative and quantitative approach. For instance, we correct the technique of a player by increasing and decreasing the range of motion in order to produce the desired effects [3]. Through the biomechanical studies and approaches, many fast

A. Hazari et al., *Conceptual Biomechanics and Kinesiology*,
https://doi.org/10.1007/978-981-16-4991-2_5

bowlers have been reinforced to improve their performance level and speed of bowling while modifying the range of motion available at the wrist, elbow, and shoulder. It has been found that a skilled baseball pitcher uses more stride range to attain the maximum pitch speed [4].

5.3 Force–Motion Principle

In order to create or modify our motion, unbalanced forces are required. Therefore, it is very evident that muscle groups that are prime movers should be targeted to achieve the desired motion. Let us focus this concept through an example below. While doing push-ups, the vertical upward force is created by muscular contraction while downward force is created by the gravitational pull. The muscles like elbow extensors, shoulder abductors, and extensors need to be trained to improve the performance, and therefore, exercises are prescribed accordingly. In simple words, the force-motion principle can be identified as targeting a specific group of muscles that drive the motion.

5.4 Force–Velocity Relationship

The force–velocity relationship states that the force produced by muscles decreases with increasing velocity of concentric actions, whereas the resistive force increases with increasing velocity of eccentric actions as shown by the graphical representation below (Fig. 5.1). Studies have shown that isolated muscles can resist twice the isometric strength during a fast eccentric contraction [5]. This is applied in heavy weight lifting. Also, this could be the basis for eccentric strength being more than the concentric.

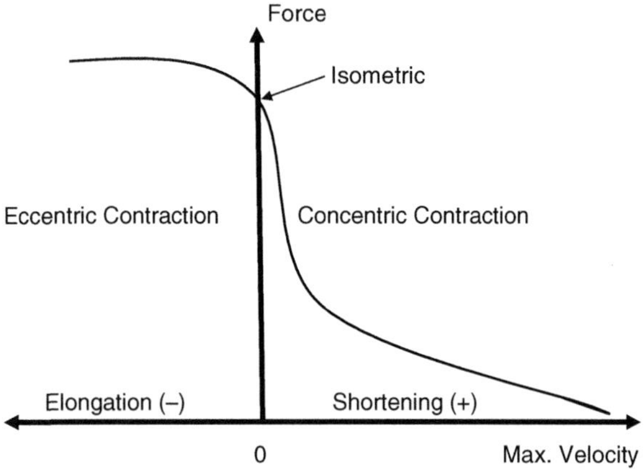

Fig. 5.1 Depicts force–velocity curve

5.4.1 Effect of Training on Force–Velocity Curve

Training can shift the direction of force velocity curve based on the principle of specificity. Strengthening with heavy weights shifts the curve upward while doing isometric slow concentric actions, whereas speed training can improve muscle forces at higher concentric speeds (Fig. 5.2).

5.5 Force–Length Relationship

In the previous chapter, we have learnt that an optimal length is required to produce maximal tension in skeletal muscles. When the muscle is shortened to produce the maximum tension, the phenomenon is known as **active insufficiency** as seen with weak wrist flexors grip strength while the wrist is complexly flexed. On the other hand, if the tension produced by the muscles is inadequate due to overstretching, it leads to **passive insufficiency.**

5.5.1 Stretch-Shortening Cycle (SSC)

Studies have shown that if a muscle contracts eccentrically with a pause and followed by a concentric contraction, the force produced is higher. This can be understood as the principle of transfer of energy. During the eccentric muscular contraction, the energy is stored as potential energy which is converted to strong kinetic energy during the concentric actions. This phenomenon is known as **stretch-shortening cycle** and frequently used in **plyometric trainings**. The concept of SSC shall be discussed in detail in the kinesiology section.

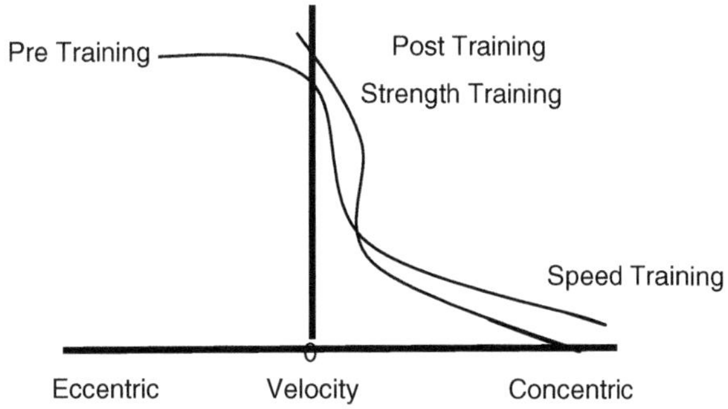

Fig. 5.2 Depicts effects of training on force–velocity curve

5.6 Summary

The principles of kinematics and kinetics are useful to understand the biomechanical function of the human body. These principles can be used in rehabilitation and sports performance training efficiently to obtain desired results.

References

1. Hudson JL. Guidelines and standards: application of concepts. Boston: Paper presented to the AAHPERD National Convention; 1989.
2. Knudson D. Fundamentals of biomechanics. Berlin: Springer Science & Business Media; 2007.
3. Knudson D, Morrison C. Qualitative analysis of human movement. 2nd ed. Champaign, IL: Human Kinetics; 2002.
4. Montgomery J, Knudson D. A method to determine stride length for baseball pitching. Appl Res Coach Athl Ann. 2002;17:75–84.
5. Alexander RM. The human machine. New York: Columbia University Press; 1992.

Lever Systems at Human Joints and Muscles

6

6.1 Introduction

A lever is a rigid bar which is capable of movement about a fixed axis called the **fulcrum (F)** [1]. The lever system in the human body plays a vital role in driving the correct kinetics and kinematics of joints. In other words, the principles of the lever system in human biomechanics allow us to be more efficient in our quality of movements. All the three types of the lever system prevail in human biomechanics, and we shall learn about them in detail here. The force applied at one point of lever is known as the **effort (E),** and the force applied at the second point is known as the **weight (W)**. The perpendicular distance from fulcrum to effort is known as the **effort arm (EA)**, and the perpendicular distance from the fulcrum to weight is known as the **weight arm (WA)**. In the human body, the following structures act as the respective parts of the lever system commonly.

Lever: Bones

Fulcrum: Joints

Efforts: Force produced by muscles

Weight: External force or resistance applied e.g; gravitational force.

6.2 Types of Lever in Human Body

6.2.1 First-Order Lever

In the first order/class of lever, the fulcrum lies between the effort and weight. Thus, there can be three possibilities: (a) fulcrum is exactly between or at midpoint where EA = WA, (b) fulcrum is shifted toward the effort where WA > EA, and (c) fulcrum is shifted toward weight where EA > WA as shown in (Fig. 6.1a, b, and c).

Now let us understand the first class of lever in human biomechanics with an example.

A. Hazari et al., *Conceptual Biomechanics and Kinesiology*, https://doi.org/10.1007/978-981-16-4991-2_6

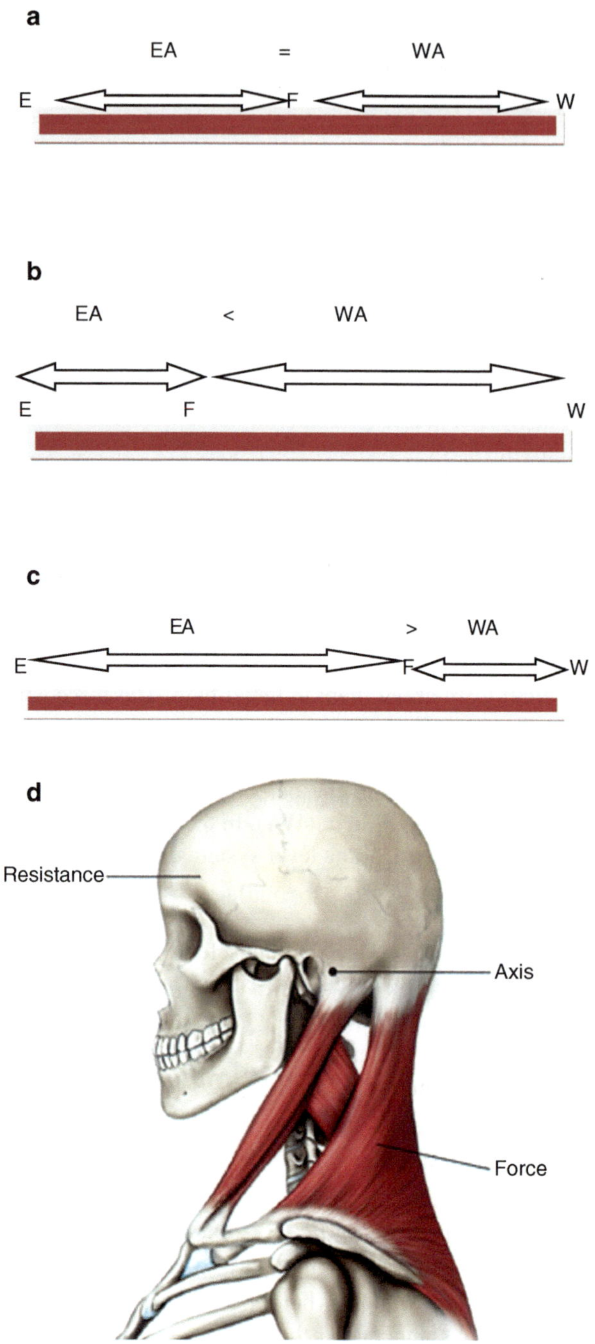

Fig. 6.1 (**a**) First-order lever where EA is equal to WA. (**b**) First-order lever where EA is smaller than WA. (**c**) First-order lever where EA is greater than WA. (**d**) Showing example of the first-class lever where the posterior neck muscle acts as effort against the weight of skull due to gravity at the atlanto-occipital joint as fulcrum

Nodding Movement of Head (Fig. 6.1d): The fulcrum is situated in between the weight and effort where

Lever: Skull

Fulcrum: Atlanto-occipital joint

Weight: gravity pulling the skull anteriority

Effort: Force produced by contraction of posterior neck muscles at the occipital bone.

6.2.2 Second-Order Lever

In the second-order/class lever system, the weight is situated between the fulcrum and effort so that the effort arm is always longer than the weight arm (Fig. 6.2a). In the event, the muscles force as effort will be more efficient since the larger perpendicular distance will demand less force to counteract the weight. Thus, the second class of lever is at **mechanical advantage (MA)**. Mathematically, the MA is the ratio of weight (W) to effort (E). In second-class levers, MA > 1 always is compared to first class where MA can be equal to, less than, or more than 1 depending upon the position of fulcrum (Fig. 6.1a, b, c). The second-class lever is also known as the **"lever of power."**

Let us see the example for the second class of lever in the human body.

Standing on toes (Fig. 6.2b): The weight lies between the fulcrum and effort where, Lever: Tarsal and metatarsal bones, Fulcrum: Metatarsophalangeal joint, Weight: Body weight transmitted through the ankle joint on the ground, Effort: Contraction of calf muscles applied at the insertion on calcaneum.

6.2.3 Third-Order Lever

In the third-order/class lever, the effort is situated between the fulcrum and weight such that the weight arm is longer than the effort arm (Fig. 6.3a). It is the most common type of lever found in the human body. Clearly, the muscles require more force to counter the weight produced by the gravity or weight segment, and thus, the third-order lever is always at mechanical disadvantage because MA < 1 always.

Let us see the example for the third class of lever in the human body.

Flexion movement at the elbow joint (Fig. 6.3b).

Lever: Forearm

Fulcrum: Elbow joint

Weight: External load in palm or gravity

Effort: Contraction of elbow flexor muscles (brachialis).

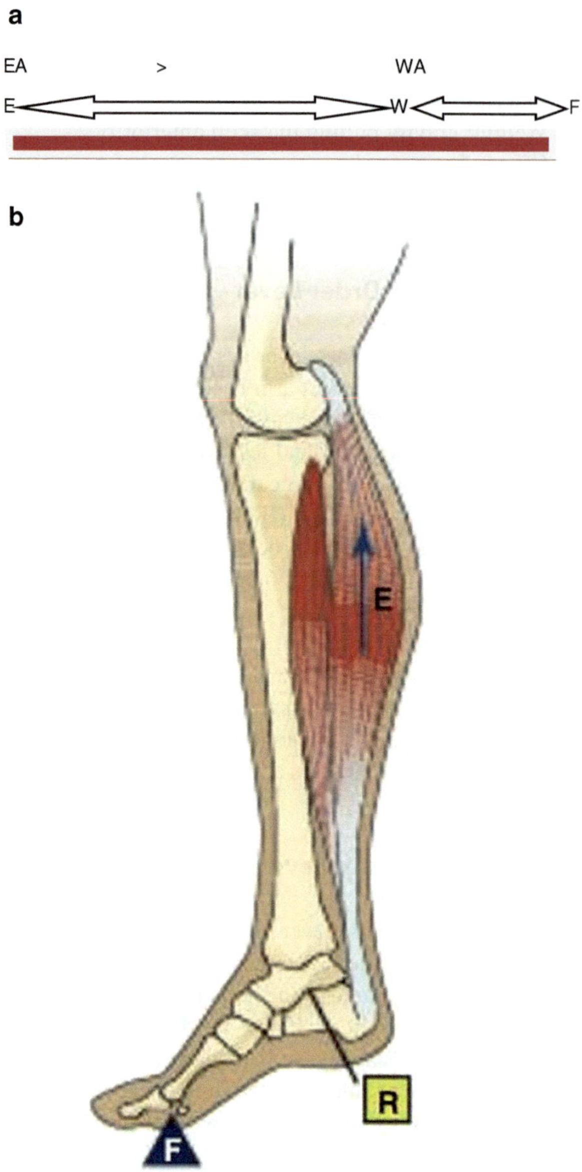

Fig. 6.2 (**a**) Second-class lever where the weight lies between the effort and fulcrum. (**b**) Showing example of the second-class lever where the calf muscle acts as effort against the weight of the foot and gravity at MTP joint as fulcrum

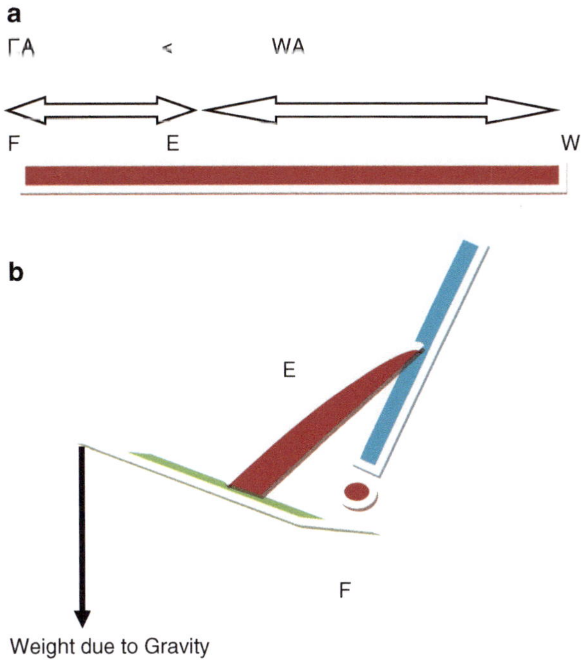

Fig. 6.3 (**a**) Third-class lever where effort lies between the weight and fulcrum. (**b**) Showing the third-class lever where the brachialis muscle is effort acting against the weight of gravity at the elbow joint as fulcrum

6.3 Summary

The lever system in the human body forms the basis of the mechanical system (kinematics and kinetics) where the muscles are considered as the effort, the joint as the fulcrum, and external resistance in the form of gravity acts as the weight. The human body is predominantly composed of the third-class lever, and thus, appropriate strength in muscles is an essential component to maintain lever and balance in the body segments.

Reference

1. Levangie PK, Norkin CC. Joint structure and function: a comprehensive analysis. Philadelphia, PA: F.A. Davis; 2011.

Kinetics and Kinematics of Temporomandibular Joint

7

7.1 Anatomical Background

The temporomandibular joint (TMJ) is the articulation between temporal bone superiorly and mandible head inferiorly (Fig. 7.1, [1]). The joint is separated by a thin disk dividing the joint into upper and lower segments. Though the TMJ lacks a defined hyaline cartilage, it is classified as the **synovial joint** where the upper segment forms a plane gliding type and lower segments act as the hinge joint [2]. Unlike other joints the TMJ is unique in its structure where both the articulating surfaces are convex, and therefore, the role of the articular disk is very well identified. The disk is biconcave in shape to allow maximum congruency (Fig. 7.1). Any dysfunction in the anatomical representation would lead to significant changes in the joint biomechanics. In this chapter, we shall concentrate on the kinematics and kinetics of the TMJ in detail.

7.2 Kinematics

The TMJ joint has the ability to move in all planes. The most common activities that reflect the TMJ motion are swallowing, chewing, grinding, yawning, etc. The major motions of TMJ are contributed by the lower segments such as mandibular motion followed by accessory gliding movements at the upper joints segments [2]. The following motions are available for TMJ.

7.2.1 Mandibular Elevation and Depression

(a) **Osteokinematics:** The TMJ joint elevation and depression can be typically understood as opening and closing of the mouth. Thus, it is a sagittal plane movement through the frontal axis (Fig. 7.2). During depression the mandible/

A. Hazari et al., *Conceptual Biomechanics and Kinesiology*,
https://doi.org/10.1007/978-981-16-4991-2_7

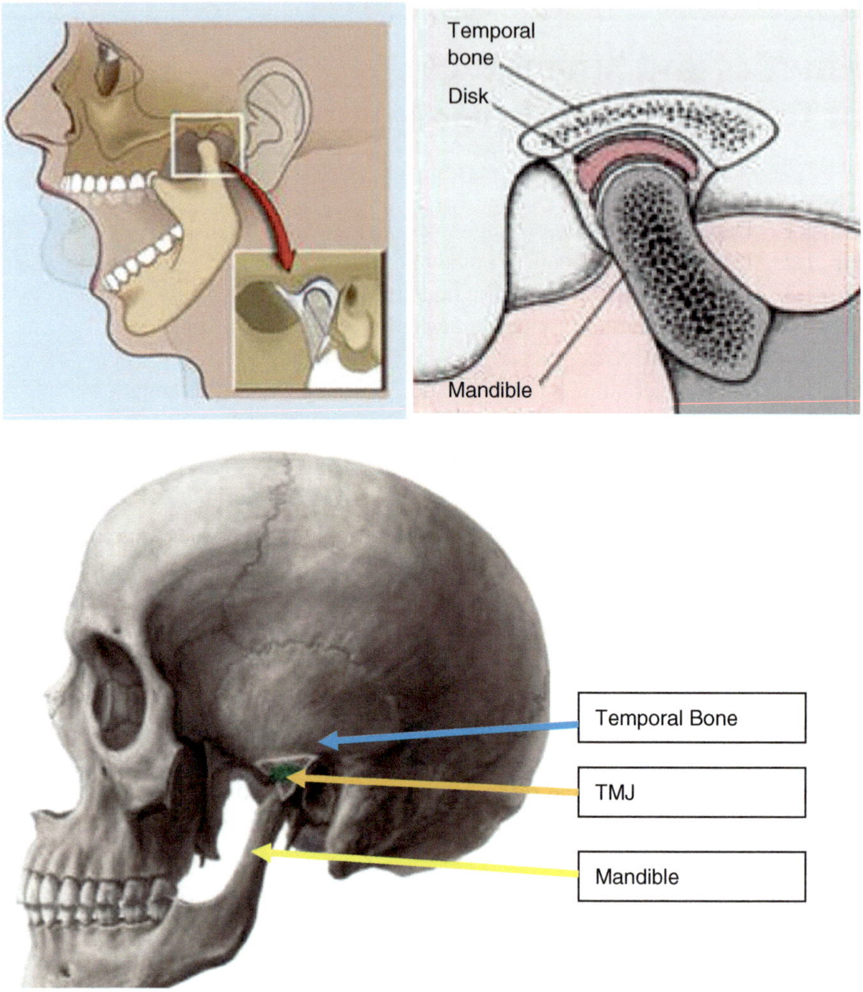

Fig. 7.1 Depicts articulating structure of TMJ

lower segment hangs down as seen in mouth opening, and during elevation the mandible pulls up from down while closing the mouth.

(b) **Arthrokinematics:** We have learnt that both the articulating surfaces are convex separated by the disk which is biconcave. Since the lower segment is mobile facilitated by motion of disk over the mandibular condyle, it can be seen as the convex on concave motion where rotation and gliding will take place in opposite directions. However, the motion of mandibular condyle takes place in the same direction at TMJ making it an exception to pure arthrokinematics rule. In the context of mouth opening/ mandibular depression, the head or condyle of the mandible would rotate anteriorly and also glide anteriorly/inferiorly

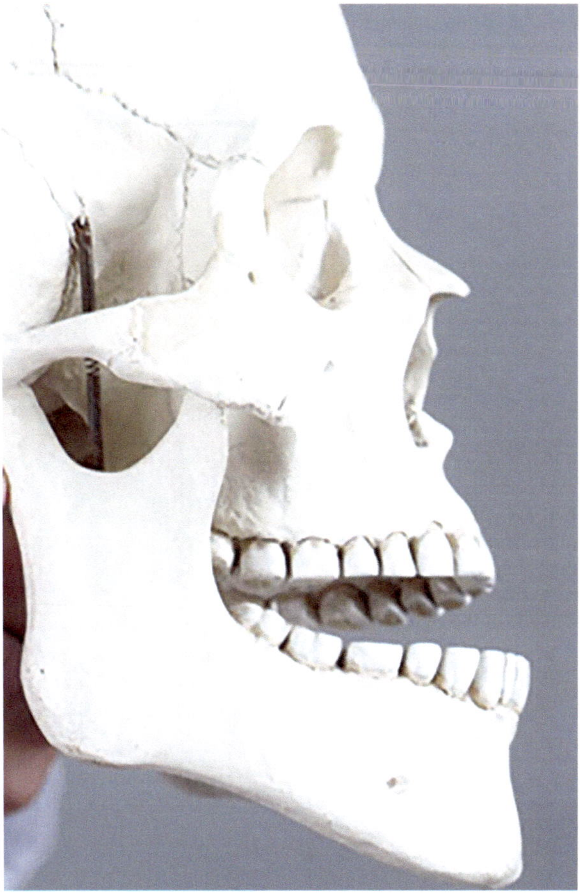

Fig. 7.2 Showing TMJ depression (mouth opening)

(Fig. 7.3). The rotation and gliding for elevation would take place in the reverse direction. The normal range of motion at TMJ is 40 to 50 mm [3]. Clinical evaluation can be done by placing three stacked fingers while asking the patient to open mouth maximally.

7.2.2 Mandibular Protrusion and Retrusion

(a) **Osteokinematics:** The mandibular protrusion and retrusion could be seen as the sagittal plane motion where pure translation takes place in forward and backward direction, respectively.
(b) **Arthrokinematics:** During protrusion there is anterior and inferior translation of disk along with mandibular condyle. The normal range of motion for protrusion is 6 to 9 mm [4]. The reverse sequence takes place for retrusion which is limited to 3 mm.

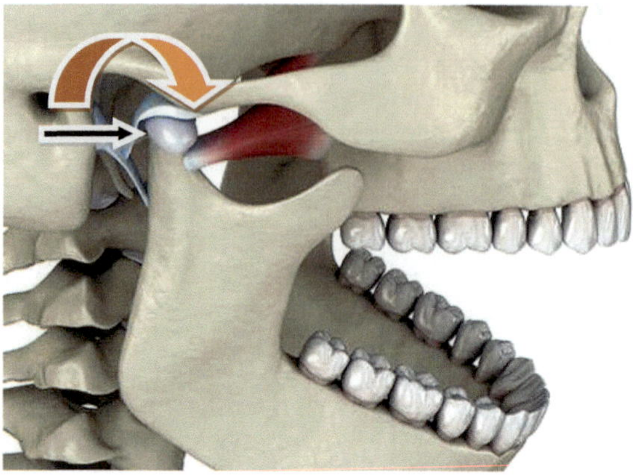

Fig. 7.3 Showing anterior rotation and translation of mandibular condyle during mouth opening

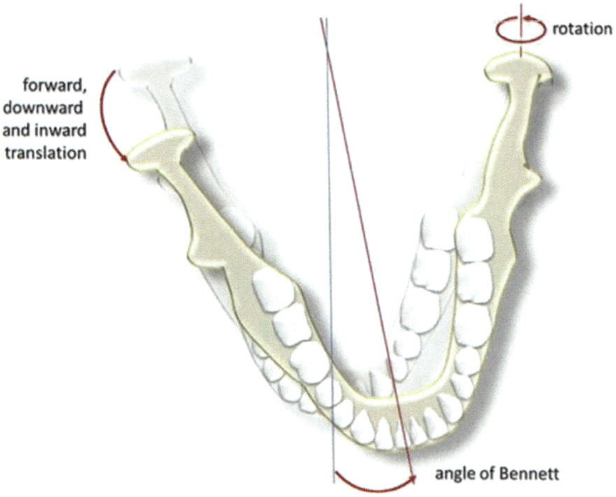

Fig. 7.4 Arthrokinematics of mandibular lateral deviation. While doing right side movement, the condyle of the right spins whereas the left side translates forward

7.2.3 Mandibular Lateral Deviation

(a) **Osteokinematics:** The TMJ lateral deviation is the combination of transverse and sagittal plane movements.

(b) **Arthrokinematics:** During deviation to one side, the ipsilateral side spins and the contralateral side translates forward (Fig. 7.4; [2]). The normal range of motion for lateral deviation is 8 mm [5].

7.3 Kinetics

The TMJ is very functional and thus stability of the joint is of utmost importance. Both active and passive stabilization is seen at TMJ by the muscles, ligaments, and capsules. The disk which enhances the congruency of the joint has additional function of stress-bearing. In the upcoming section, we shall learn about the muscles and ligament in detail which help to determine the kinetics of the TMJ.

7.3.1 Kinetics at TMJ Depression

The driving force for TMJ motion is mainly controlled by the muscles of the jaw. Ligaments and other soft tissues have accessory functions. Soft tissues that drive the motion would be called as the **facilitators** and those which check/stabilize the excessive motion would be referred to as the **limiters** in this book.

Facilitators:

(a) Internal Factor:
Muscles – **Digastric and lateral pterygoid** [6]

(b) External Factor: Gravity

7.3.2 Kinetics at TMJ Elevation

Facilitators: **temporalis muscle, masseter muscle, internal pterygoid muscle** [7].

7.3.3 Kinetics at TMJ Protrusion

Bilateral action of the **masseter, internal pterygoid** [8], and **external pterygoid** muscles [9].

7.3.4 Kinetic at Retrusion

Bilateral action of the posterior fibers of the **temporalis** muscles with assistance from the anterior portion of the **digastric muscle** [10].

7.3.5 Kinetics at Lateral Deviation

The **internal and external pterygoid** muscles are responsible for contralateral deviation whereas the temporalis helps in ipsilateral deviation. The **temporalis** muscle can deviate the mandible to the same side [10].

7.4 Pathomechanics at TMJ

In this section, we shall discuss about the most common pathology at TMJ which alters the normal biomechanics. As a result changes in motion (kinematics) and force distribution (kinetics) could be seen which could further damage the structure and function.

7.4.1 Internal Derangement and Inflammatory Conditions

One of the most common *pathologies* due to lax soft tissues and aging, e.g., **bruxism, osteoarthritis, capsular fibrosis, and hypermobility.**

7.4.2 Altered Kinetic and Kinematic Chain

The TMJ dysfunction can be often correlated to the altered closed kinematic chain either from the proximal segment (altered cervical curvature) or distal segment (altered foot arches). The muscle attached at TMJ also has its attachment at the cervical region and therefore weakness or shortening of muscles at neck can cause biomechanical dysfunctions at the TMJ. A recent study established a close relationship between the TMJ and cervical spine where correction of one could lead to improvement in the other segment [11]. Similarly pronated foot has been associated with TMJ pain [12].

7.5 Summary

The TMJ joint is unique in its biomechanical function where the muscular force (kinetics) drives the motion (kinematics) in all planes. It has six degrees of freedom where rotation and gliding do not follow the normal arthrokinematic rule (Table 7.1).

Table 7.1 Kinematic and kinetics at TMJ

Available motion	Osteokinematic	Arthrokinematic (condyle over disk)	Range of motion	Muscles
Depression	Sagittal plane	Anterior rotation and inferior glide	0 to 40 mm	Digastric and external pterygoid
Elevation	Sagittal plane	Posterior rotation and superior glide	40 to 0 mm	Temporalis, masseter, internal pterygoid muscle
Protrusion	Sagittal plane	Anterior and inferior translation	0 to 9 mm	Masseter, internal pterygoid, external pterygoid
Retrusion	Sagittal plane	Posterior and superior translation	0 to 3 mm	Temporalis digastric muscle
Lateral deviation (right)	Transverse plane	Spin at right and forward translation at left	0 to 8 mm	Internal and external pterygoid (left) Temporalis (right)
Lateral deviation (left)	Transverse plane	Spin at left and forward translation at right	0 to 8 mm	Internal and external pterygoid (right) Temporalis (left)

References

1. Lintonbon D. The osteopathic approach to the temporomandibular joint (TMJ). Positive Health Online, 128; 2006.
2. Levangie PK, Norkin CC. Joint structure and function: a comprehensive analysis. Philadelphia, PA: F.A. Davis; 2011.
3. Hertling D, Kessler RM. Management of common musculoskeletal disorders: physical therapy principles and methods. Philadelphia, PA: Lippincott Williams & Wilkins; 2006.
4. Kraus SL, editor. TMJ disorders: management of the craniomandibular complex, vol. 18. New York: Churchill Livingstone; 1988.
5. Kraus S. Temporomandibular joint. In: Saunders H, Saunders R, editors. Evaluation, treatment, and prevention of musculoskeletal disorders. 2nd ed. New York: Viking Press; 1985.
6. Matsumoto MAN, Matsumoto W, Bolognese AM. Study of the signs and symptoms of temporomandibular dysfunction in individuals with normal occlusion and malocclusion. Cranio. 2002;20(4):274–81.
7. Yang X, Pernu H, Pyhtinen J, Tiilikainen PA, Oikarinen KS, Raustia AM. MRI findings concerning the lateral pterygoid muscle in patients with symptomatic TMJ hypermobility. Cranio. 2001;19(4):260–8.
8. Kraus S. Temporomandibular joint. In: Saunders H, Saunders R, editors. Evaluation, treatment, and prevention of musculoskeletal disorders. 3rd ed. Bloomington, MN: Educational Opportunities; 1993.
9. Murray GM, Orfanos T, Chan JY, et al. Electromyographic activity of the human lateral pterygoid muscle during contralateral and protrusive jaw movements. Arch Oral Biol. 1999;44:269–85.
10. Williams P. Gray's Anatomy. 38th ed. New York: Churchill Livingstone; 1999.
11. Walczyńska-Dragon K, Baron S, Nitecka-Buchta A, Tkacz E. Correlation between TMD and cervical spine pain and mobility: is the whole body balance TMJ related? BioMed Res Int. 2014;2014:582414.
12. Saito ET, Akashi PMH, Sacco IDCN. Global body posture evaluation in patients with temporomandibular joint disorder. Clinics. 2009;64(1):35–9.

Kinematics and Kinetics of Shoulder Complex, Elbow, and Wrist

8

8.1 Introduction

The shoulder complex biomechanics is one of the most dynamic areas of study and research. This chapter focuses on the clinical significance of shoulder joint complex. It is important to understand the biomechanics to precisely analyze the complexity of the joint [1]. The shoulder complex includes four joints linked to each other functionally and structurally, viz. **(i). glenohumeral (GH) joint, (ii). acromioclavicular (AC) joint, (iii). sternoclavicular (SC) joint,** and **(iv). scapulothoracic (ST) joint** (Fig. 8.1). The shoulder muscles act as a team to create highly organized motion. The weakening in muscle alters the shoulder joint's natural kinematic chain [2]. Since the shoulder joint is a complex anatomical structure with multiple landmarks, it would be helpful to briefly focus on the anatomy of the shoulder joint before we discuss the kinematics and kinetics in detail.

8.1.1 Anatomical Background

The osseous or bony segments in the shoulder joint include the sternum, clavicle, scapula, and humerus.

Clavicle

The clavicle is located between the sternum and the scapula, from which it holds the upper body to the humerus. The S-shaped clavicle acts as the osseous linkage between the shoulders to the thorax. It serves as a solid brace for the upper limb located away from the thorax and enhances more comprehensive position for usage of the limb. As other long bones, both medial and lateral epiphysis are found in the clavicle [3].

A. Hazari et al., *Conceptual Biomechanics and Kinesiology*,
https://doi.org/10.1007/978-981-16-4991-2_8

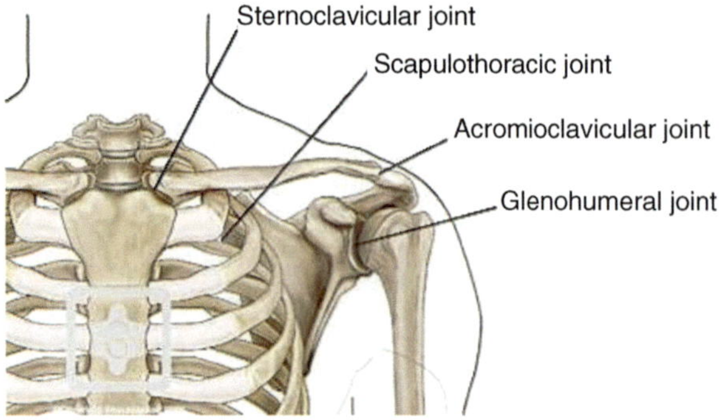

Fig. 8.1 Depicts the shoulder joint complex

Scapula

The scapula, also commonly known as the **shoulder blade**, is an agile, thin, flat trilateral bone located on the posterolateral side of the thorax. It consists of **two surfaces, three borders, three angles, and three processes.** The subscapular fossa is the somewhat concave in the anterior part of the bone, which allows the scapula to slide efficiently over the posterior convex thoracic cage. The concave glenoid fossa is a partially oval-shaped surface that articulates with the humeral head to constitute the glenohumeral joint [4]. The borders of superior and inferior glenoid tubercles and the glenoid fossae act as proximal attachments for the long head of biceps and the long head of triceps, respectively. The posterior part of the scapula is divided by the scapular spine into the supraspinatus fossa and the infraspinatus fossa. A long, flattened bony prominence from the most superior-lateral part of the scapula is the acromion process. The finger-like projection of bone from the anterior surface of the scapula, palpable about 1 inch below the most concave component of the distal clavicle, is the coracoid process. At the inferior angle, or tip, of the scapula, the medial and lateral edges of the scapula meet. The inferior angle is clinically significant in helping to detect scapular motion [5].

Humerus

The humerus is one of the long bone in the human body. At the shoulder joint, the humerus binds across the glenoid fossa of the scapula to the axial body and attaches to the ulna's olecranon fossa at the elbow.

The upper part of the humerus is the center for many ligament and muscle attachments. The humeral head is half of a complete sphere of glenoid cavity that articulates with glenoid fossa. A sharp, anterior bone projection just beneath the humeral head is the lesser tubercle. The greater tubercle creates a more oval lateral bony projection. Deltoid tuberosity located at lateral upper one-third of humerus shaft allows insertion for all deltoid muscle. The radial or spiral groove, which aids in establishing the insertion of the triceps' lateral and medial heads, runs obliquely across the humerus' posterior surface [6].

Sternum

The sternum is a flat bone located at the midpoint of the anterior thorax that consists of the **manubrium, the body, and the xiphoid process**. The manubrium is the upper section of the sternum that articulates with the upper part of the second costal cartilage, the first rib of both sides, and the clavicle that forms the sternoclavicular joint. The quadrangular manubrium sits between third and fourth thoracic vertebrae level. The jugular notch is the thickest portion of the bone where the anterior side is convex, and the posterior side is concave. The sternum's body has margins that articulate with the cartilage of the second to seventh rib. The "sword-shaped" xiphoid process is the lower tip and the shortest portion of the sternum. It does not articulate with the ribs. The xiphoid process anchors many essential muscles, such as the rectus abdominis and the transversus thoracis, and the diaphragm that is essential for breathing [7].

8.1.2 Articulations

The four main articulations involved with the shoulder complex, including the clavicle, sternum, ribs, scapula, and humerus, work in association with each other for providing a large range of movements in all three planes of motion to the upper extremity. The shoulder complex includes 3 physiological joints and 1 functional joint. The interplay among four articulations of the shoulder complex results in a synchronized pattern of arm movement. The movements involved are continuous and interdependent at each joint, although they occur at different intervals and at different angles of motion of the shoulder complex. GH, AC, and SC joints link the upper extremity to the axial skeleton. The ST joint makes the scapula to glide over the contours of the thoracic posterior wall. To attain normal shoulder girdle movements, all four joints work together cohesively. The shoulder joint movement is complex and allows dynamic relationship between ligament constraints, muscle forces, and bony articulations. The GH joint is primarily designed for mobility and ability to move and position the hand through a wide range in space, permitting the largest range of motion in the body [5].

8.1.3 Shoulder Joint Angulations

The humerus head faces medially, superiorly, and posteriorly, whereas the glenoid fossa is 6–7 degrees retroverted from scapular plane. Thus, at the shoulder joint two distinct angulations are observed.

Angle of inclination—the angle is formed in the frontal plane between the line intersecting the humerus shaft and humerus neck. The normal angle ranges from 130 to 150° (Fig. 8.2).

Angle of torsion—the angle is formed in the transverse plane between the line intersecting the humerus neck and humerus condyles. The normal angle ranges from 20 to 30° (Fig. 8.2).

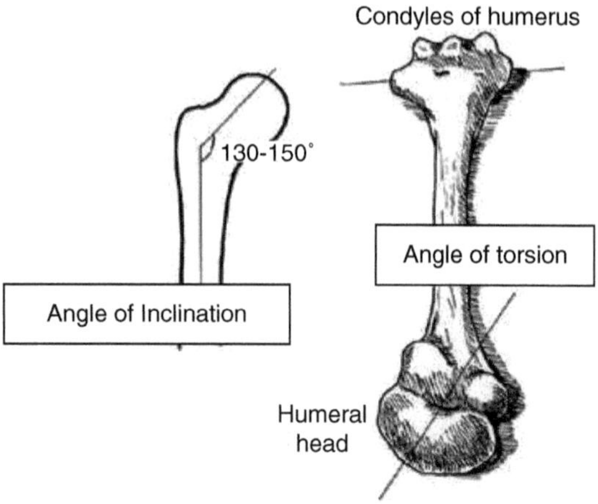

Fig. 8.2 Depicts shoulder angle of inclination and angle of torsion

8.1.4 Supporting Structures of the Shoulder Joint

(A) **Joint Capsule**

The joint capsule is essential passive stabilizer of the shoulder joint. The glenohumeral joint capsule is thickened at the front and is double the diameter of the humerus head. It supplies much of its extensibility before and below, and "winds up" during abduction with external rotation. The joint capsule has an inherent negative intra-articular pressure, which holds the joint together.

(B) **Ligaments**

I. **Glenohumeral ligaments**

(a) Superior GH ligament restricts the external rotation and inferior translation of the humeral head.

(b) Middle GH ligament restricts external rotation and forward translation at 45° abduction [8].

(c) The lower GH ligament is the thickest of the ligaments and checks superior, anterior, and posterior humeral motion as it consists of three distinct portions: the anterior and posterior band and the axillary pouch.

II. **Coracohumeral ligament:** It covers the superior glenohumeral joint forward and upward and fills the space between supraspinatus and subscapularis tendons. It limits forward and downward translation at lower levels of elevation, extension, and extension with adduction.

III. **Coracoacromial ligament:** It is a strong triangular band, extending from coracoid process to acromion that together forms a vault for the protection of the head of the humerus [9].

IV. **Capsular ligament (fibrous capsule):** It is made up of collagen and elastic fibers. Medially it is attached to the peripheral margin of the glenoid cavity outside of the glenoid labrum. Common shoulder pain is seen at the sub-acromial space, which includes the theoretic space between the coracoacromial arch and the humerus head. More precisely, the sub-acromial canal lies underneath the acromion, the coracoid process, and the coracoacromial ligament. Sub-acromial bursa situated here provides lubrication for rotator cuff (RC) tendons, the insertion of the long head of the biceps [10].

(C) **Glenoid Labrum**

It is a ridge-like fibrocartilaginous, connective tissue that enhances the articular surface of the humeral head with the glenoid fossa. It provides the main link to the glenohumeral ligaments and provides partial origin to the long head of biceps tendon, capsule, and anatomical neck of the humerus. Its curvature enhances about 50% of the depth of the shoulder joint. During shoulder joint rotational movement, it stretches out at the front with external rotation and stretches out at the back with internal rotation. Therefore, the lack of labrum integrity has been shown to minimize translation resistance of shoulder complex about 20% [11].

(D) **Bursae**

Bursa is a fluid-filled synovial sac that serves as cushion for tendons and other structures of joint. In the shoulder, there are eight bursae. The bursae have both a nerve supply and a mechanoreceptor that serves to provide proprioceptive information on the joint position of the shoulder. This indicates that bursae do not act exclusively as a lubricator between tissues. The upper limits of the bursae are the coracoacromial ligament, the bone of the acromion, and the deltoid. The lower limits are the humeral head, the joint of the shoulder, and the supraspinatus [12].

8.2 Kinematics of the Shoulder Joint Complex

8.2.1 Sternoclavicular Joint: Osteokinematics and Arthrokinematics

It is a plain synovial saddle joint formed between the sternum and the clavicle. The joint is slightly incongruent and maintained by the fibrocartilagenous disk. It allows 3 degrees of rotatory and 3 degrees of translatory motions.

(A) **Elevation and Depression:** Sagittal axis motion with convex clavicle moving over concave manubrium. Following convex–concave rule, the clavicle rolls superiorly and glides inferiorly during elevation and vice versa during depression.

Available Range of Motion: Elevation—48 degree, Depression—15 degree.

Ligaments

Costoclavicular ligaments: strong ligament between the clavicle and the first rib.

Function—checks elevation of the lateral end of the clavicle.

Interclavicular ligament

Function—limits upward glide of the medial clavicle and depression of the distal clavicle.

(B) **Protraction and Retraction**

It is a vertical axis motion of the concave clavicle over the convex manubrium. During protraction, the clavicle rolls and slides anterior and vice versa for retraction following concave–convex rule.

Available Range of Motion: Protraction – 15–20 degrees, Retraction – 20–30 degrees.

Ligaments

Sternoclavicular ligaments: Anterior and posterior bands. The function is to check anterior–posterior translation of medial end of the clavicle.

(C) **Anterior and Posterior Rotation**

It occurs as a spin along the long axis of the clavicle from its neutral position.

Available Range of Motion: Anterior rotation—10 degrees, Posterior rotation—50 degrees.

8.2.2 Acromioclavicular Joint

It is a plane synovial joint articulation between a larger lateral clavicle end and a small facet on acromion process of the scapula, thus making it very incongruent. There are three major motions available.

(A) **Internal and External Rotation**

It takes place in the vertical axis and transverse plane. During internal rotation, the glenoid fossa comes anterior and medial and just the opposite for external rotation.

Available Range of Motion—combined 30 degrees.

(B) **Anterior and Posterior Tipping**

It is a coronal axis motion. During anterior tipping, acromion tips forward with the inferior angle of the scapula tipping backward. During posterior tipping, the inferior angle of scapula tips forward.

Available Range of Motion: Anterior tipping—30 degrees, Posterior tipping – 40 degrees.

(C) **Upward and Downward Rotations**

Sagittal axis and frontal plane motion.

Available Range of Motion: Upward rotation—30 degrees, downward rotation—17 degrees

Ligaments

Acromioclavicular ligaments—Superior and inferior fibers. Provides anterior and posterior stability

Coracoclavicular ligament—Two portions

Lateral trapezoid portion—it is horizontally oriented and restricts posterior translation of distal clavicle.

Medial conoid portion—it is vertically oriented and acts as restraint for superior–inferior translation.

They together also limit upward rotation of the scapula on the thorax and play the most critical role in posterior rotation of the clavicle for efficient external rotation during arm abduction beyond 90 degrees.

8.2.3 Scapulothoracic Joint

It is a functional joint rather than a true anatomical joint as it lacks the fibrocartilage union and synovial tissue [8]. The motion at ST joint is a resultant motion at AC and/or SC both [8].

(A) **Elevation /depression:** translatory cephalocaudal motion of the scapula on the thorax accompanied by:
SC JOINT—elevation /depression
AC JOINT—internal/external rotation

(B) **Protraction/retraction:** translatory motion toward or away from the vertebral column accompanied by:
SC JOINT—protraction/retraction
AC JOINT—internal/external rotation

(C) **Anterior/posterior tipping:** predominant AC joint motion along with anterior/posterior rotation of the clavicle at SC joint.

8.2.4 Glenohumeral Joint

The glenohumeral joint (GH joint) is a true synovial ball-and-socket joint that is responsible for integrating the upper limb to the thorax, allowing 6 degrees of freedom. The articulation is formed by the humeral head and the glenoid fossa of the scapula. It is to be noted that the glenoid fossa is relatively flat, in comparison to a large, rounded hemisphere of the head of the humerus. In addition, it is approximated that only one-quarter of the humeral head articulates with the glenoid fossa during movements at any given time; thus, the joint is mobile but incongruent [13].

The shoulder joint is the most moveable and least stable joint in the body and commonly gets dislocated. The joint itself has limited bony congruency, due to the relatively large surface area of the humeral head in correspondence with the fossa, and consequently relies heavily on surrounding soft tissues for support. In a healthy individual, the surrounding passive structures (labrum, joint capsule, and ligaments) along with active structures (muscles and related tendons) work together to maintain static and dynamic stability throughout movement. It is to note that the ligaments and capsules of the GH joint are relatively thin and even provide secondary joint stability, while the rotator cuff muscles provide the primary stabilizing force along surrounding musculature [14]. The shoulder joint has motion available in all planes as listed below.

(A) **Flexion and Extension**

Osteokinematics: frontal axis and sagittal plane motion.

Arthrokinematics: in open chain kinematics where the humeral head (convex) would move over the glenoid fossa (concave), rolling and gliding would occur in opposite directions. Thus for flexion, rolling would be anterior and gliding would be posterior. The extension of the shoulder would consist of opposite direction of rolling and gliding. In close kinematic chain, where the humerus is fixed and glenoid fossa moves, rolling and gliding would occur in the same directions. Thus, the flexion would consist of anterior rolling and anterior gliding and opposite for extension.

(B) **Adduction and Abduction**

Osteokinematics: sagittal axis and frontal plane motion.

Arthrokinematics: in open chain kinematics for abduction, rolling would be superior and gliding would be inferior. The adduction of the shoulder would consist of opposite direction of rolling and gliding.

(C) **Medial and Lateral Rotation**

Osteokinematics: vertical axis and transverse plane motion.

Arthrokinematics: in open chain kinematics for lateral rotation, rolling would be backward and gliding would be forward. The internal rotation of the shoulder would consist of opposite direction of rolling and gliding.

(D) **Circumduction:** Apart from the pure planar motions, the shoulder has the ability to show multiplanar motions completing a circular motion (combination of flexion, adduction, abduction, and extension in succession) and thus known as the circumduction of the shoulder.

The open chain kinematics of the glenohumeral joint has been summarized in Table 8.1.

Table 8.1 Open chain kinematic characteristics of the glenohumeral joint

Available motion	Osteokinematics Axis and plane	Arthrokinematics Convex–concave	Range of motion in degrees
Flexion	Frontal axis and sagittal plane	Rolling—anterior Gliding – posterior	0–110
Extension	Frontal axis and sagittal plane	Rolling—posterior Gliding—anterior	0–60
Adduction	Sagittal axis and frontal plane	Rolling—inferior Gliding—superior	120–0
Abduction	Sagittal axis and frontal plane	Rolling—superior Gliding—inferior	0–120
Medial rotation	Vertical axis and transverse plane	Rolling—forward Gliding—backward	0–90
Lateral rotation	Vertical axis and transverse plane	Rolling—backward Gliding—forward	0–90

8.2.5 Scapulohumeral Rhythm

During arm elevation, the scapulohumeral rhythm is defined as the ratio of the glenohumeral movement to the scapulothoracic movement. This is most often determined by taking the total amount of shoulder elevation by the scapular upward rotation (humerothoracic) (scapula-thoracic). The shoulder's actions are paired with the scapula's actions. This increases the upper extremity's available range of motion and allows the glenoid fossa to be positioned in a more stable position relative to the humeral head. Scapulohumeral rhythm, first published by Codman in the 1930s, is the kinematic interaction between the scapula and the humerus [15]. For the optimal function of the shoulder, this interaction is substantial. In the sagittal plane, the scapulohumeral rhythm or ratio is significantly greater in **scapular plane** (less scapular movement and more humeral movement) than in other planes. The dominant side, consistent with the findings, shows significantly higher scapulohumeral rhythm values than the non-dominant side, but only in the coronal and scapular planes. Scapulohumeral rhythm is described as a proportion of humeral elevation and scapulothoracic rotation. The overall ratio of 2:1 during arm elevation is commonly used. According to the 2-to-1 ratio framework, flexion or abduction of 90° in relation to the thorax would be accomplished through approximately 60° of GH and 30° of ST motion [16]. When there is a change in the normal position of the humerus-related scapula, this can cause scapulohumeral rhythm dysfunction, often referred to as reverse scapulohumeral rhythm and lead to **scapular dyskinesia**.

8.3 Kinetics of the Shoulder Joint: Static and Dynamic Stability of the Shoulder Complex

The shoulder joint stability depends on the coordination of both static (noncontractile, for example, ligaments) and dynamic (contractile, for example, tendons and muscles) tissues [17].

8.3.1 Static Stability

Articular Surfaces and Bony Geometry

The bony architecture of the glenohumeral joint is favorable but loses bony stability to excessive joint mobility. It is thicker at the periphery and provides the basis for the effect of rotator cuff muscles [18]. The superior angulation of the glenoid helps to hold the humeral head stuck over it and the joint capsule created a negative joint pressure to maintain the integrity of the humeral head within the labrum.

Static Structures and Mechanoreceptors

The shoulder complex's static structures, which include the labrum (a fibrocartilaginous ring), capsule, cartilage, ligaments, and fascia, serve collectively as the physical constraints of the bone and provide the shallow glenoid fossa with a deepening effect. They also provide protection in addition to their passive stabilization function through the different mechanoreceptors embedded within their fibers. Mechanoreceptors are distinguished by specialized nerve endings that are responsive to tissue mechanical deformation and thus contribute to the regulation of the adjacent muscles' motor responses. Our sense of touch, sensation, and proprioceptive positioning is the responsibility of mechanotendinous receptors (muscle spindles and golgi tendon organs), capsuloligamentous receptors (Ruffini and Pacinian corpuscles), and cutaneous receptors (Meissner, Merkel, and free nerve endings), as well as providing feedback on muscle length, stress, and orientation, further to the speed and strength of the contractions of the muscle fibers. Through feed-forward and feedback input, the passive shoulder structures provide a neural defense mechanism that directly mediates reflex musculature stabilization around the glenohumeral joint [19].

8.3.2 Dynamic Stability

For the shoulder complex, dynamic stabilization results in a large range of mobility and provides stability within the functional limit. The active, contractile tissues constitute the dynamic stabilizers of the shoulder complex. This includes the muscles of the rotator cuff, the deltoid muscles, and the scapular muscles that regulate the scapulohumeral rhythm and are linked to proprioception-related sensorimotor system (joint location sense, kinesthesia (sense of motion), sense of force, sense of vibration, and sense of velocity). The dynamic stability of the shoulder is mainly dependent on the following muscles [8].

Deltoid and glenohumeral stabilization

Rotator cuff and glenohumeral stabilization

Supraspinatus stabilization

Long head of biceps brachii

Role of Deltoid—The dynamic stabilizers must function in an effective synergistic fashion for optimum shoulder stabilization. During movement, these dynamic stabilizers help maintain the central position of the humeral head within the glenoid

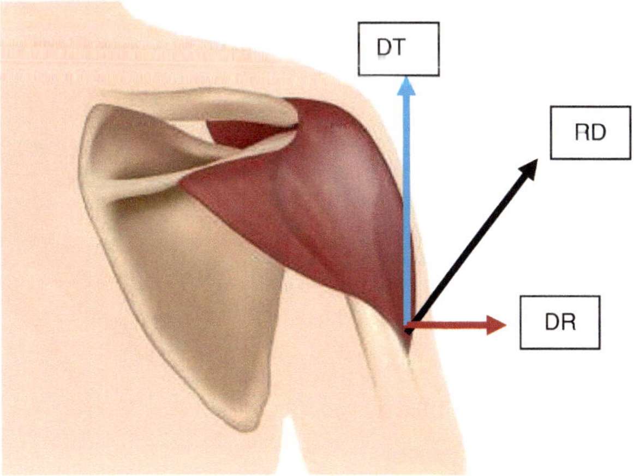

Fig. 8.3 Deltoid muscle action in dynamic stability

fossa. The deltoid muscle is the shoulder joint's powerful abductor and elevator. Let us analyze the deltoid force vector in dynamic stabilization. Consider Fig. 8.3, where the deltoid muscle is shown for the right shoulder from the posterior aspect.

The resultant force of the deltoid is in the line of pull for middle deltoid muscle as shown by the force vector RD. The force vector DT represents the translatory force of the deltoid, whereas the force vector DR represents the rotatory force component of deltoid. It can be seen that the translatory component of the deltoid helps to counter the gravitation pull on the humerus. Also, the rotatory component helps to abduct the humerus.

Role of Infraspinatus, Teres Minor, and Subscapularis Muscles

It is essential that during overhead activities, three rotator cuff muscles (infraspinatus, teres minor, and subscapularis supraspinatus) activate the humeral head in synergy with the deltoid for dynamic stability [20]. Let us understand the force action of rotator cuff muscle like infraspinatus, teres minor, and subscapularis using Fig. 8.4 for the right shoulder in posterior view. The resultant line of action for infraspinatus and teres minor is shown by the force vector R where the rotatory component is represented by RR and translatory component by RT. It can be seen that the translatory component of these rotator cuffs is offsetting the excessive superior translation of deltoid as it is directed downward against the superior translation of deltoid. Also, the rotatory component pushes the head of the humerus into the glenoid fossa medially and thus helps to maintain the congruency during motion.

Role of Supraspinatus—The supraspinatus muscle has a significant role in dynamic stabilization of shoulder during abduction. Referring to Fig. 8.5, we see that the small translatory component of the supraspinatus is upward represented by force vector ST, and thus assists the deltoid to overcome gravitational pull downward. In contrast, the rotatory component is large and directed to push the humerus head into the glenoid fossa, thus maintaining stability. The larger rotatory component allows the supraspinatus muscle to initiate the abduction.

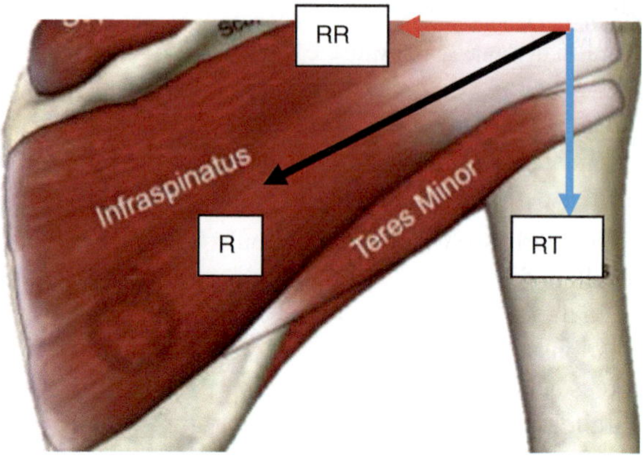

Fig. 8.4 Depicts force vector for rotator cuff in dynamic shoulder stabilization

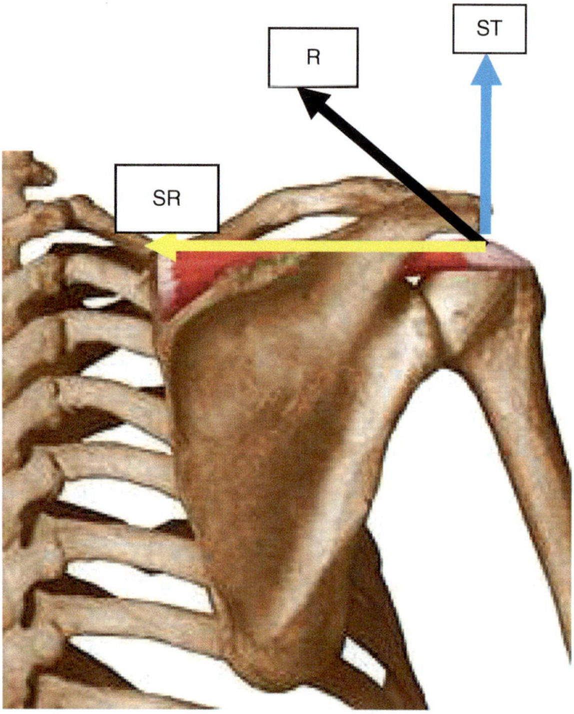

Fig. 8.5 Depicts force vector for supraspinatus in dynamic shoulder stabilization

Biceps Mechanism

It has been found that both long heads of biceps have parallel functions as anterior stabilizers of the glenohumeral joint during actions like abduction and external rotation, and their role increases as shoulder stability declines. Both heads of the biceps act as stabilizing function in resisting anterior humeral head displacement, and consideration is given to strengthening the biceps during rehabilitation programs in case of chronic anterior instability [21].

Secondary Dynamic Stabilizers

Teres major is a small muscle located at the lateral border of the scapula. It forms the inferior border for both the triangular space and quadrangular space. It is sometimes called **"lat's little helper"** because of its synergistic action along with latissimus dorsi.

Latissimus dorsi, which means "broadest muscle of the back," is one of the widest muscles in the human body. Also known as the "lat's," it is a very thin triangular muscle that is not used strenuously in common daily activities but plays a crucial role in many pelvic and core exercises and swimming.

Pectoralis major is a thick, fan-shaped muscle, situated at the chest which makes up the bulk of the chest muscles [22].

8.3.3 Force Couple

A force couple is described as two equal but oppositely directed forces acting concurrently on the opposite sides of an axis that generates rotation. In biomechanics, a couple is a system of muscular forces that results in a moment without a resultant force and creates a rotation without translation at joint.

Synergetic mechanisms of co-contractions of the shoulder muscles, appropriate positioning, control, and coordination of the shoulder, as well as the scapula-thoracic complex obtain dynamic stabilization during upper extremity movements. The dynamic stability of the shoulder complex can be categorized into the following: glenohumeral stability (local) and scapulothoracic stability (global). Deltoid muscle plays an important role as a stabilizer in glenohumeral joint stability and is generally accepted during abduction as a prime mover for the glenohumeral joint, along with the supraspinatus muscle [23]. The deltoid is the main muscle that is responsible for abduction from 15 to 90 degrees of the limb. It also acts as a humeral head stabilizer, especially in cases of carrying a load. The supraspinatus muscle is better than the other three rotator cuff muscles in terms of the location. **The deltoid and rotator cuff** muscles for the motions associated with the glenohumeral joint work as a mutual force coupling. Biomechanical alterations can be caused by an imbalance of one or more of these muscles and contribute to shoulder disorders such as impingement syndromes, bursitis, instability, scapular dyskinesia, or chronic conditions associated with pathological wear and tear.

The **trapezius** muscle helps to rotate the scapula upward with **serratus anterior**, which helps maintain sub-acromial space. Serratus anterior's strong action as a protractor/upward rotator requires an opposite force to regulate this movement (equally strong antagonist). As a primary force coupling, the serratus anterior and trapezius (middle) muscles function to rotate the scapula upward [24].

8.3.4 Scapular Muscles

The scapular stability depends on the coordinated activity of the 18 muscles directly attached to the scapula. For the humeral head to remain centered and allow arm movement to occur, the scapular muscles must dynamically control the positioning of the glenoid. When there is weakness or neuromuscular dysfunction of the scapular musculature, normal scapular arthrokinematics are altered and an individual is eventually predisposed to an injury to the GH joint [25].

There is an impressive array of muscles that attach and act on the four joints of the shoulder complex as listed below.

Intrinsic muscles, commonly called as the scapulohumeral muscular group, are deeper muscles that originate from the scapula and /or the clavicle and insert on the humerus.

• Supraspinatus •Infraspinatus •Subscapularis •Teres minor.

Extrinsic muscles are larger, more superficial muscles that originate on the thorax and attach to the bones of the shoulder complex (the humerus, clavicle, and scapula).

• Latissimus dorsi • Teres major • Pectoralis major • Pectoralis minor

Other muscles act on the shoulder joint, which act as secondary movers. They are generally in the pectoral area of the body or the upper arm.

• Biceps brachii • Triceps brachii

8.4 Pathomechanics of Shoulder Complex

Freedom of movement has been produced at the cost of stability (commonly referred to as the trade-off of mobility-stability), and increasing demands for both mobility and stability, combined with complex structural and functional architecture, make the shoulder complex highly susceptible to dysfunction and instability [26].

Rotator cuff disorders: The two foremost disorders of the rotator cuff are impingement and tendinopathy. Multifactorial and pathological changes, both of which influence the quality of tendons, are the result of rotator cuff injuries and tendon degeneration. The supraspinatus, followed by infraspinatus, subscapularis, and teres minor, is commonly injured tendon of the rotator cuff muscles. The tendon tear of the subscapularis can be associated with the rupture of a biceps tendon from the bicipital tendon groove moving medially into the tendon of the subscapularis. The absence of the biceps tendon in the empty bicipital groove is confirmation of intra-articular tendon tears. Musculoskeletal pain patients often develop local tissue changes that affect functional participation, focal tenderness, and pain related to activity. This presents with pain, altered range of motion in contexts of capsuloligamentous restrictions and movement abnormalities, or altered muscle activation patterns at glenohumeral and scapulothoracic joints. The cuff accounts for a substantial share of the shoulder's rotational strength [27].

Shoulder Instability

The term "shoulder instability" refers to the inability of the glenoid fossa to retain the humeral head. A balanced joint reaction force is created by the ligamentous and muscle structures around the glenohumeral joint under normal conditions. If the integrity of any of these structures is disturbed, atraumatic or traumatic instability can result. Traumatic injury mechanisms may result in frank dislocations where there has been a loss of integrity of the joint. Instability, regardless of the mechanism of injury, can occur anteriorly, posteriorly, or in multiple directions. Traumatic shoulder instability is a common condition which is associated with high recurrence rates, particularly in young patients.

Anterior dislocation due to trauma is the most common type among the different types of joint instability, representing more than 90% of the cases. A subclassification of glenohumeral joint instability is atraumatic (non-traumatic) shoulder instability, encompassing those for whom trauma is not considered the primary etiology [28].

Shoulder Subluxation

Glenohumeral subluxation is defined as a partial or incomplete dislocation, which is usually caused by changes in the joint's mechanical integrity. The humeral head slips out of the glenoid cavity in subluxation because of weakness in the rotator cuff or a knock to the area of the shoulder. One of three types of subluxation may occur: anterior (forward), posterior (backward), and inferior (downward). The distinction is the fact that the humeral head pops back into its socket with a shoulder dislocation. Because it is the most frequently dislocated joint, the proposition that stability must be sacrificed to obtain mobility gives an effective demonstration. The glenohumeral ligament is the largest ligament, and in shoulder joint subluxation it is commonly damaged or overstretched. In individuals with hemiplegic stroke or with a paralyzed upper extremity, shoulder subluxations often occur. The incidence reported varies greatly, from 17 percent to 81 percent. Traumatic shoulder subluxation can occur in many sports, such as football, rugby, wrestling, and boxing [29].

Bicep Tendinopathy

The bicep tendinopathy pathology was extensively investigated by, among others, De Palma and Callery, Lippmann, and Hitchcock and Bechtol in the 1940s and 1950s. Neer described long head of biceps involvement in rotator cuff disease but overemphasized the role of the tendon as a humeral head depressor. Long head of biceps (LHB) tendinitis is an inflammatory tenosynovitis that occurs as the tendon travels along its constrained path within the humerus' bicipital groove. LHB tendinitis is present with anterior shoulder pain and is often exacerbated by overuse, like other types of biceps tendinopathy [30].

Labral Injuries

In different patient populations, various types of shoulder labral tears can be found. SLAP lesions are most common.

Adhesive Capsulitis

Adhesive shoulder capsulitis is a common condition that affects 2–5% of the general adult population and up to 20% of diabetes patients. The term "adhesive capsulitis" was given by Neviaser in 1945 for painful shoulder stiffening. To classify a painful shoulder condition of insidious onset with stiffness and difficulty sleeping on the affected side, Codman used the term **"frozen shoulder"**. It was Duplay, however, who originally defined the condition as "periarthrite scapulo-humerale" in 1872. Such terms are used synonymously now [31]. The deposition of hydroxyapatite crystals into the muscle tendon is the cause of the frozen shoulder. The most common site of hydroxyapatite calcification in the human body is the supraspinatus tendon. Diabetes mellitus is associated with frozen shoulder, but can also be linked with coronary artery disease, cerebral vascular disease, rheumatoid arthritis, and thyroid disease. Emig et al. found that the capsule and synovium thickness > 4 mm adjacent to the axillary recess was highly specific to adhesive capsulitis; changes in the coracohumeral ligament were found to be not consistent. Others have reported that a more common finding was scarring within the rotator interval [32].

Scapular Dyskinesia

The imbalance in the scapular muscle strength could lead to abnormal scapulohumeral rhythm as well as altered scapular tracking over the thorax which is commonly termed as scapular dyskinesis. It is thus important to know the normal resting position of scapula on the thorax as described below.

Resting position of scapula:

- Resting between 2 and 7 ribs
- 2 inches from midline
- 30–45 degrees internal rotation
- 10–20 degrees anterior tipping
- 10–20 degree of upward rotation

8.5 Biomechanics of Elbow Joint

8.5.1 Introduction

The elbow and wrist is biomechanically linked to the shoulder joint for providing a greater range of mobility through space. The elbow joint serves as the mechanical linkage between the shoulder joint and wrist and provides greater reach, whereas the wrist joint's main function is to perform intricate skilled movements. In this chapter, we shall focus on the most important aspects of elbow, wrist, and biomechanics.

8.5.2 Biomechanics of Elbow Joint

The structure and function of the elbow joint is efficient to perform both mobility and stability. The elbow joint complex is formed by the two articulations. The proximal articulation is seen between the humerus and the ulna known as the **humeroulnar** joint, as well as between the humerus and the radius known as the **humeroradial** joint. The distal articulation is formed between the radius and the ulna known as the **superior and inferior radioulnar joint.**

The humeroulnar joint is formed by the concave trochlear notch of the ulna and the convex medial humeral epicondyle. The humeroradial joint is formed by the concave radial head and convex humeral capitulum.

8.5.2.1 Osteokinematics

The predominant motion seen at the elbow joint (humeroulnar and humeroradial) is **flexion and extension** occurring in the sagittal plane and frontal axis. Minimal abduction and adduction can also be seen. The other major motion available at the superior and inferior radioulnar joint is **supination and pronation** which takes place in the transverse plane as rotational movements.

8.5.2.2 Arthokinematics

At the proximal articulation (humeroulnar and humeroradial joints), the concave surfaces of the ulna and radius move over the convex humerus in open chain kinematics. Thus, rolling and gliding would be occurring in the same direction. For instance, the elbow flexion would consist of anterior rolling and anterior gliding.

The superior radioulnar joint consists of **ulnar radial notch,** the capitulum of the humerus, and the head of the radius covered by the annular ligament. The circular ring of the annular ligament allows the movement of radius as a pure spin at the superior radioulnar joint. During pronation, the radius rides over the ulna through its spinning movement, and during supination, it comes back to resting position through reverse spin [8].

The inferior radioulnar joint is composed of the **ulnar notch** of the radius, the **articular disk (triangular fibrocartilage, TFC)**, and the **head of the ulna.** During pronation and supination, the concave surface of the ulnar notch of the radius slides around the ulnar head, and the disk follows the radius by twisting at its apex and sweeping along beneath the ulnar head [8].

Summary of Pronation and Supination Movement at Elbow

- Pronation of the forearm occurs as a result of the radius's crossing over the ulna at the superior radioulnar joint.
- During pronation and supination, the rim of the head of the radius spins within the osteoligamentous enclosure formed by the radial notch and the annular ligament.
- At the same time, the surface of the head spins on the capitulum of the humerus.
- At the distal radioulnar joint, the concave surface of the ulnar notch of the radius slides around the ulnar head.

8.5.3 Kinetics of the Elbow Joint

The elbow joint is considered to be a congruent joint because of greater articulating surface through a wider range of joint motion. The stability is provided by the joint bony structure and alignment, its surrounding ligaments and joint capsule, as well as muscles around the joint.

8.5.3.1 Ligaments of Elbow

The elbow joint is supported by a number of ligaments that provide significant static stability to the joint. The important ligaments are listed below along with their role in specific directional stability.

Annular
Quadrate ligaments
Oblique cord
Dorsal and palmar radioulnar ligament
Interosseous membrane

Medial Stability

Major stability to the medial aspect of the joint is given by the ulnar collateral ligament.

Lateral Stability

Major stability to the lateral aspect of the joint is given by the radial collateral annular ligament and lateral ulnar collateral ligament.

Proximal Radioulnar Joint

Proximal radioulnar joint is reinforced by the annular and quadrate ligaments, oblique cord (limits supination), and interosseous membrane [8, 33, 34].

Distal Radioulnar Joint

Distal radioulnar joint is stabilized by interosseous membrane [35], dorsal radioulnar ligament (limits pronation) [36], palmar radioulnar ligament (limits supination), triangular fibrocartilage, and joint capsule [8].

8.5.3.2 Muscles of the Elbow Joint

***Primary Flexors*:**

Brachialis
Brachioradialis
Biceps brachii
Accessory Flexors:
Extensor carpi radialis longus
Pronator teres
Primary Extensor:
Triceps
Anconeus
Supination and Pronation:
Supinator and Pronator Teres

8.5.3.3 Dynamic Stabilization through Muscles

The muscles of the elbow joint play a significant role in providing dynamic stabilization. The important muscles and their role have been listed below.

Triceps: a major stabilizer for flexion and extension movements. The reciprocal concentric and eccentric contraction of the triceps during push-up and push-down phase can be seen as the most significant contribution for dynamic stabilization [8].

Pronator quadratus: is active throughout supination and pronation and provides dynamic stabilization for the distal radioulnar joint [8].

Extensor carpi ulnaris: exerts a depressive force on the dorsal aspect of the ulnar head during supination [8]. Extensor carpi radialis brevis also provides support for the forearm, for gripping during pronation.

Pronators teres: are found to be most efficient around the neutral position of the forearm when the elbow is flexed to 90 degree [8].

8.5.3.4 Pathomechanics

The elbow joint is prone to multiple pathological changes as a number of structures pass through the joint. The most commonly seen pathomechanics due to specific dysfunction has been listed here and few important ones from the biomechanical perspective have been discussed in detail.

(a) **Osseous and articular changes:**
Bone hypertrophy
Traction spur formation
Osteochondral defects
Loose bodies
Joint degeneration
Chondromalacia
Osteophyte formation
Epiphyseal and apophyseal deformity
Dislocation/avulsion fractures

(b) **Ligamentous changes:**
Sprains
Calcium deposition

(c) **Soft tissue changes:**
Synovitis
Adhesive capsulitis

(d) **Tendon related:**
Tendinosis/itis

(e) **Muscle alteration:**
Myositis fibrosis/ossification

(f) **Nerve related:**
Ulnar nerve entrapment
Pronator syndrome
Radial nerve entrapment

Lateral antebrachial cutaneous nerve entrapment

(g) **Bursae:**
Olecranon bursitis

(h) **Altered angulations:**
Cubitus valgus and varus

Lateral Epicondylitis

The term **"tennis elbow"** was introduced by Major and Runje in 1880. Nirschl and Pettrone concluded tendinitis as a misnomer as no inflammatory cells were present in the surgical specimens. Pathoanatomical changes primarily occur in extensor carpii radialis brevis and secondarily to extensor digitorum. Therefore, it is commonly referred to as lateral epicondylitis due to its site as common extensor origin. "Backhand strokes" are the most common predisposing factors with repetitive muscle contraction.

Medial Epicondylitis

Similar pathological condition with respect to histological pathoanatomy. It frequently occurs over the tip of the medial epicondyle, and extending distally 1 to 2 inches along the common flexor origin. Muscles like pronator teres and flexor carpii radialis are usually involved.

Little League Elbow

A pathology seen at younger age groups due to an accelerated apophyseal growth and delayed medial epicondyle epiphysis growth. It is often characterized by weakness in triceps giving a sense of *"locking"* or *"catching"* sensation with decreased pronation and supination range.

Cubitus Valgus and Cubitus Varus Angle

The long axis of humerus forms a medial angle with the long axis of ulnar in full extended position at the elbow. The normal angle is 15° [8]. If the angle increased significantly, it is referred to as cubitus valgus, whereas a significant decrease in angle is known as the cubitus varus.

8.6 Biomechanics of the Wrist Joint

The wrist is the most distal segment of the upper extremity which performs many intricate skilled movements. The main function of the wrist is to allow grasping power and precision. The wrist joint complex is formed by the radiocarpal and midcarpal articulations. As the name suggests, the radiocarpal joint is articulation between the distal end of the radius with the carpal bones and midcarpal joint is the articulation between the proximal and distal rows of the carpal bones. Clearly, there is no bony articulation between the ulna and carpal though ligamentous connections are present [8].

8.6.1 Proximal Articulation

The proximal articulation of the wrist joint complex is formed by the radiocarpal joint consisting of **radius, radioulnar disk, and triangular fibrocartilage complex (TFCC)** proximally with the **scaphoid, lunate, and triquetrum carpal bones** distally [8]. The distal radius is biconcave which articulates with the scaphoid and lunate predominately; the triquetrum is connected through TFCC [8]. The **pisiform, anatomically** part of the proximal row, does not participate in the radiocarpal articulation, and its role is to increase the moment arm of the flexor carpi ulnaris muscle [8].

8.6.2 Distal Articulation

The distal articulation of the wrist joint complex is formed by the midcarpal joint consisting of scaphoid, lunate, and triquetrum as proximal carpal bones and **trapezium**, **trapezoid, capitate**, and hamate as distal carpal bones [8].

8.6.3 Angulations of the Wrist

The radiocarpal joint has two main angulations. One consists of radial inclination in the frontal plane of about 23° (Fig. 8.6), and the other in the sagittal plane of about 11 ° [8]. These angulations add to the functional anatomy of the wrist.

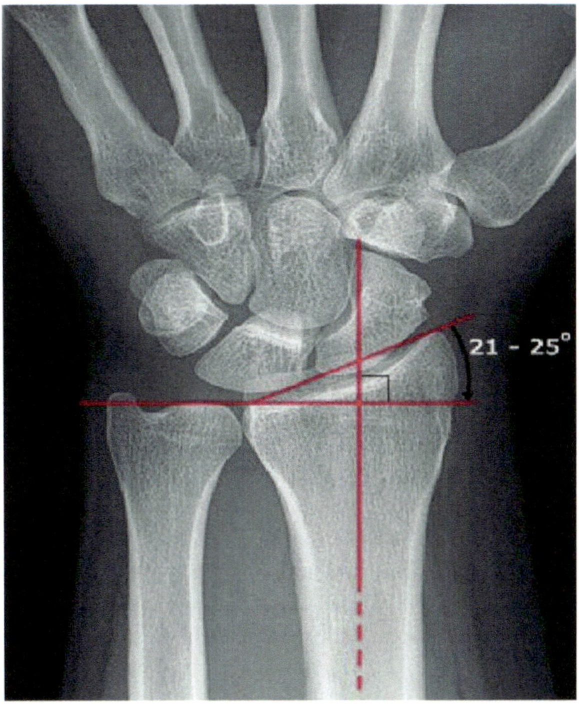

Fig. 8.6 Radiographic presentation for inclination of radius

The length of the ulna compared to the distal radius has a significant role in biomechanical function [37]. If the ulna is significantly shorter at the distal radial end, then it is known as **negative ulnar variance** and if longer then **positive ulnar variance** [8]. These conditions affect the range of wrist joint motion significantly.

8.6.4 Ligaments of the Wrist Complex

A number of ligaments are present at the wrist to support the numerous structures passing in and around. They primarily work to stabilize the carpals bone during dynamic motion. The ligaments of the wrist can be broadly classified as the intrinsic and extrinsic ligaments [8].

Intrinsic ligaments: ligaments that interconnect the carpals themselves and are also known as intercarpal or interosseous ligaments.

Extrinsic ligaments: ligaments that connect the carpals to the radius or ulna proximally or to the metacarpals distally.

Both intrinsic and extrinsic ligaments are found on the volar and dorsum of the hand and have been named accordingly as listed below.

Radiocarpal Ligaments: Extrinsic on the Volar Aspect

Radioscaphocapitate ligament: connects radius proximally and scaphoid and capitates distally

Radioscapholunate ligament: connects radius proximally and scaphoid and lunate distally

Radiolunate ligament: connects radius proximally and lunate distally

Radiocarpal Ligaments: Intrinsic on the Volar Aspect

Scapholunate ligament: connects scaphoid to lunate

Lunotriquetral ligament: connects lunate to triquetrum

Radiocarpal Ligaments: Dorsal Aspect

Dorsal radiocarpal ligament

Dorsal intercarpal ligament

Ulnocarpal Ligament: Volar Aspect

Ulnar collateral ligament

Ulnolunate ligament

8.6.5 Kinematics of the Wrist Joint Complex

The kinematics of the wrist joint is facilitated by its anatomical structure, presence of the carpal bones, and active and passive forces caused by the muscles and ligaments, respectively. The wrist can be considered to have four degrees of freedom predominately seen as flexion/extension as well as radial/ulnar deviation [8].

8.6.5.1 Osteokinematics

The flexion and extension movement of the wrist is seen in the sagittal plane through the frontal axis. The radial and ulna deviation occurs in the frontal plane and sagittal axis for a human body in anatomical position.

8.6.5.2 Arthokinematics

The flexion and extension of the wrist can be described as proposed through a sequence of events [38]. The event has been step wise listed for wrist extension from full flexion and shown in Fig. 8.7. The opposite sequence takes place for flexion from extension.

Radial and ulnar deviation: The proximal carpal row displays a unique "reciprocal" motion with radial and ulnar deviation [39]. In radial deviation, the carpals slide ulnarly on the radius. The carpal motion not only produces deviation of the proximal and distal carpals radially, but simultaneous flexion of the proximal carpals and extension of the distal carpals [8, 40]. Opposite motion can be seen in ulnar deviation.

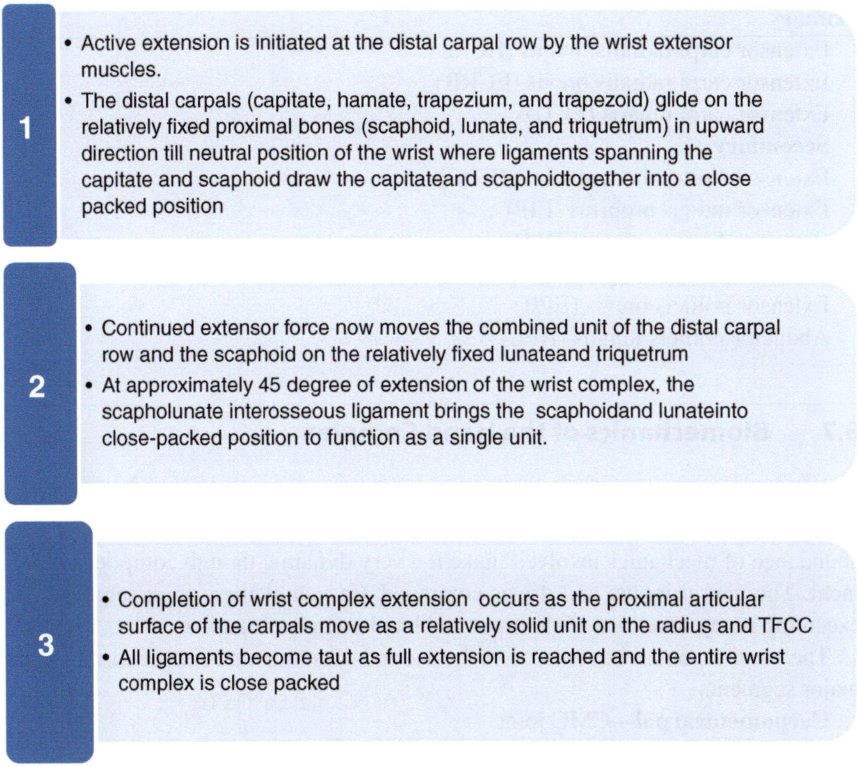

Fig. 8.7 Flowchart depicting sequence of events for extension of wrist from full flexion

8.6.6 Kinetics of the Wrist Joint

The wrist joint is well supported by its ligaments passively and muscle actively. We have learnt about the ligaments of the wrist in the above section and muscles of the wrist complex have been listed here. The force generated by the muscles of the wrist is used for gripping function at hand which we shall deal in the next section.

Wrist Flexors
Primary
Palmaris longus (PL)
Flexor carpi radialis (FCR)
Flexor carpi ulnaris (FCU)
Secondary
Flexor digitorum superficialis (FDS)
Flexor digitorum profundus (FDP)
Flexor pollicis longus (FPL)

Wrist Extensors
Primary
Extensor carpi radialis longus (ECRL)
Extensor carpi radialis brevis (ECRB)
Extensor carpi ulnaris (ECU)
Secondary
Extensor digitorum communis (EDC)
Extensor indicis proprius (EIP)
Extensor digiti minimi (EDM)
Extensor pollicis longus (EPL)
Extensor pollicis brevis (EPB)
Abductor pollicis longus (APL)

8.7 Biomechanics of the Hand Complex

The biomechanics of the wrist complex varies at the most distal segment and thus discussed separately here under the hand complex. The smaller structures with abundance of mechanics involved make it a very dynamic though complicated segment. The neural inputs and drives command are met at these joint segments for executing many functions of the wrist and hand complex together.

The hand complex consists of 19 bones and 19 joints and classified into three major segments.

Carpometacarpal—CMC joint
Metacarpophalangeal—MP joint
Interphalangeal joint—IP joints, the proximal (PIP) and distal (DIP)

8.7.1 CMC Articulation

The second metacarpal articulates with trapezoid primarily and secondarily with the trapezium and capitate.

The third metacarpal articulates primarily with the capitate.

The fourth metacarpal articulates with the capitate and hamate.

The fifth metacarpal articulates with the hamate.

8.7.2 Kinematics at CMC

Flexion and extension movement is predominately seen for the second and third CMC allowing one degree of freedom. The fifth CMC has two degrees of freedom [8].

8.7.3 Kinematics at MCP

Each of the four MCP joints of the fingers is composed of the convex metacarpal head proximally and the concave base of the first phalanx distally. The MCP joint is condyloid with two degrees of freedom: flexion/extension and abduction/adduction [8].

8.7.4 Kinematics and Kinetics of the Hand Complex

The hand complex as a whole has major kinematic function, i.e., gripping and grasping, collectively known as the **prehension.** The presence of hand arches, ligaments, and volar plates makes it suitable structure to perform this function.

Prehension can be categorized as either **power grip** (full-hand prehension) or **precision handling** (finger-thumb prehension) [8]. Power grip is generally a forceful act resulting in flexion at all finger joints. When the thumb is used, it acts as a stabilizer. Precision is the skillful placement of an object between fingers or between finger and thumb [8].

8.7.4.1 Functional Steps of Power Grip [8]

1. Opening the hand
2. Positioning the fingers
3. Bringing the fingers to the object
4. Maintaining a static phase that actually constitutes the grip

8.7.4.2 Types of Power Grip

Cylindrical grip: In the dynamic phase of finger closing, the power is generated by the flexor digitorum muscle (FDP). In the static phase, the flexor digitorum superficialis (FDS) muscle assists when the intensity of the grip requires greater force [8].

The interossei muscles function primarily as MP joint flexors and abductors/adductors.

Cylindrical grip is typically performed with the wrist in neutral flexion/extension and slight ulnar deviation [8], e.g., holding a mobile phone tightly.

Spherical grip: The main distinction from the cylindrical grip can be made by the greater spread of the fingers to encompass the object. This demands more interosseous activity, e.g., holding a ball.

Hook grip: Finger flexors are mostly active at the DIP and PIP. It does not involve thumb, e.g., holding a vegetable bag.

Lateral Prehension

Lateral prehension is seen when contact occurs between two adjacent fingers.

The MP and IP joints are usually maintained in extension and MP joints simultaneously abduct and adduct, e.g., holding a paper between the fingers.

8.7.4.3 Functional Steps of Precision Grip [8]

It shares the first three steps of the power grip sequence but does not contain a static phase. In precision handling, the fingers and thumb grasp the object for the purpose of manipulating it within the hand.

8.7.4.4 Types of Precision Grip

Pad-to-pad: involves opposition of the thumb pad to the pad of the finger**.** The MP and PIP joints of the fingers are partially flexed, whereas the DIP joint may be fully extended or in slight flexion, e.g., holding a pen.

Tip-to-tip: the IP joints of the finger and thumb are flexed. The MP joint is ulnarly deviated (with finger tip pointed radially) to present the tip of the finger to the thumb, e.g., pinching movements.

Pad-to-side: involves contact between pad of the thumb and side of the index finger. Thumb is more adducted and less rotated as compared to tip-to-tip and requires higher activity of **flexor pollicis brevis and opponens pollicis**, e.g., holding a key.

8.8 Functional Position of Wrist and Hand

At wrist: slight extension (20°) and slight ulnar deviation (10°).

At hand: fingers moderately flexed at the MP joints (45°) and PIP joints (30°) and slightly flexed at the DIP joints.

8.9 Pathomechanics of Wrist and Hand

Wrist Instability

Dorsal intercalated segmental instability (**DISI**)—extended lunate. Progressive degenerative problem from an untreated DISI is known as **scapholunate advanced collapse (SLAC wrist).**

Volar intercalated segmental instability (VISI)—flexion of lunate and scaphoid and extension of triquetrum and distal carpal row.

Kienbock's disease—avascular necrosis of the lunate associated with negative ulnar variance.

8.10 Summary

The shoulder joint complex is the most mobile segment of the human body. The joint complex has integrated function with the scapula and thorax. The stability in shoulder is compromised over mobility, and thus, utmost care should be taken to maintain the normal biomechanics during dynamic motions in particular.

The elbow, wrist, and hand complex are the anatomical units of the upper limb which serve the stability and mobility functions. These add to the wider domain of biomechanics for the shoulder. The dysfunction of any segment would cause altered biomechanics and interfere with smooth functioning of the upper extremity. Due to the close kinematic chain function, the joints are interlinked and can affect the adjoining segments significantly.

References

1. Veeger HEJ, Van Der Helm FCT. Shoulder function: the perfect compromise between mobility and stability. J Biomech. 2007;40(10):2119–29.
2. Guyot J. The shoulder girdle and shoulder joints. In: Atlas of human limb joints. Berlin, Heidelberg: Springer; 1981. p. 102–33.
3. Rockwood CA, editor. The shoulder, vol. 1. Philadelphia, PA: Elsevier Health Sciences; 2009.
4. Calais-Germain B. Anatomy of movement. Seattle: Eastland Press; 1993.
5. Kent BE. Functional anatomy of the shoulder complex: a review. Phys Ther. 1971;51(8):867–88.
6. Terry GC, Chopp TM. Functional anatomy of the shoulder. J Athl Train. 2000;35(3):248.
7. O'Neal ML, Dwornik JJ, Ganey TM, Ogden JA. Postnatal development of the human sternum. J Pediatr Orthop. 1998;18(3):398–405.
8. Levangie, P.K. and Norkin, C.C., 2011. Joint structure and function: a comprehensive analysis.
9. Warner JJ, Bowen MK, Deng X, Torzilli PA, Warren RF. Effect of joint compression on inferior stability of the glenohumeral joint. J Shoulder Elb Surg. 1999;8(1):31–6.
10. Werner CM, Nyffeler RW, Jacob HA, Gerber C. The effect of capsular tightening on humeral head translations. J Orthop Res. 2004;22(1):194–201.
11. Provencher CMT, Bhatia S, Ghodadra NS, Grumet RC, Bach BR Jr, Dewing LCB, et al. Recurrent shoulder instability: current concepts for evaluation and management of glenoid bone loss. JBJS. 2010;92(Supplement_2):133–51.
12. Mohammed H, Skalski MR, Patel DB, Tomasian A, Schein AJ, White EA, et al. Coracoid process: the lighthouse of the shoulder. Radiographics. 2016;36(7):2084–101.
13. Schenkman M, Rugo de Cartaya V. Kinesiology of the shoulder complex. J Orthop Sports Phys Ther. 1987;8(9):438–50.
14. Wilk KE, Arrigo CA, Andrews JR. Current concepts: the stabilizing structures of the glenohumeral joint. J Orthop Sports Phys Ther. 1997;25(6):364–79.

15. Lee ST, Moon J, Lee SH, Cho KH, Im SH, Kim M, Min K. Changes in activation of serratus anterior, trapezius and latissimus dorsi with slouched posture. Ann Rehabil Med. 2016;40(2):318.
16. Sugamoto K, Harada T, Machida A, Inui H, Miyamoto T, Takeuchi E, et al. Scapulohumeral rhythm: relationship between motion velocity and rhythm. Clin Orthop Rela Res®. 2002;401:119–24.
17. Lugo R, Kung P, Ma CB. Shoulder biomechanics. Eur J Radiol. 2008;68(1):16–24.
18. Voight ML, Cook G. Clinical application of closed kinetic chain exercise. J Sport Rehabil. 1996;5(1):25–44.
19. Janwantanakul P, Magarey ME, Jones MA, Dansie BR. Variation in shoulder position sense at mid and extreme range of motion. Arch Phys Med Rehabil. 2001;82(6):840–4.
20. Lephart SM, Pincivero DM, Giraido JL, Fu FH. The role of proprioception in the management and rehabilitation of athletic injuries. Am J Sports Med. 1997;25(1):130–7.
21. Itoi EIJI, Kuechle DK, Newman SR, Morrey BF, An KN. Stabilising function of the biceps in stable and unstable shoulders. J Bone Joint Surg Br. 1993;75(4):546–50.
22. Sharkey NA, Marder RA, Hanson PB. The entire rotator cuff contributes to elevation of the arm. J Orthop Res. 1994;12(5):699–708.
23. Baritello O, Khajooei M, Engel T, Kopinski S, Quarmby A, Mueller S, Mayer F. Neuromuscular shoulder activity during exercises with different combinations of stable and unstable weight mass. BMC Sports Sci Med Rehabil. 2020;12(1):1–14.
24. Phadke V, Camargo PR, Ludewig PM. Scapular and rotator cuff muscle activity during arm elevation: a review of normal function and alterations with shoulder impingement. Braz J Phys Ther. 2009;13(1):1–9.
25. Fayad F, Roby-Brami A, Yazbeck C, Hanneton S, Lefevre-Colau MM, Gautheron V, et al. Three-dimensional scapular kinematics and scapulohumeral rhythm in patients with glenohumeral osteoarthritis or frozen shoulder. J Biomech. 2008;41(2):326–32.
26. Happee R, Van der Helm FCT. The control of shoulder muscles during goal directed movements, an inverse dynamic analysis. J Biomech. 1995;28(10):1179–91.
27. Haza T. Analysis of the scapulohumeral rhythm. Nihon Seikeigeka Gakkai Zasshi. 1988;62(12):1105–19.
28. Gebremariam L, Hay EM, van der Sande R, Rinkel WD, Koes BW, Huisstede BM. Subacromial impingement syndrome—effectiveness of physiotherapy and manual therapy. Br J Sports Med. 2014;48(16):1202–8.
29. Warren RF. Subluxation of the shoulder in athletes. Clin Sports Med. 1983;2(2):339–54.
30. Uhorchak JM, Arciero RA, Huggard D, Taylor DC. Recurrent shoulder instability after open reconstruction in athletes involved in collision and contact sports. Am J Sports Med. 2000;28(6):794–9.
31. Kim DS, Yoon YS, Yi CH. Prevalence comparison of accompanying lesions between primary and recurrent anterior dislocation in the shoulder. Am J Sports Med. 2010;38(10):2071–6.
32. Shah N, Lewis M. Shoulder adhesive capsulitis: systematic review of randomised trials using multiple corticosteroid injections. Br J Gen Pract. 2007;57(541):662–7.
33. Fuss F. The ulnar collateral ligament of the human elbow joint. Anatomy, function and biomechanics. J Anat. 1991;175:203.
34. Olsen BS, Vaesel MT, Sojbjerg JO, et al. Lateral collateral ligament of the elbow joint: Anatomy and kinematics. J Shoulder Elbow Surg. 1996;5:103.
35. Murray WM, Buchanan TS, Delp SL. The isometric functional capacity of muscles that cross the elbow. J Biomech. 2000;33:943.
36. Savnik A, Jensen B, Nørregaard J, Egund N, Danneskiold-Samsøe B, Bliddal H. Magnetic resonance imaging in the evaluation of treatment response of lateral epicondylitis of the elbow. Eur Radiol. 2004;14(6):964–9.
37. O'Driscoll SW, Farsetti P, Potenza V, Caterini R, Ippolito E. Stress radiographs are important in diagnosing valgus instability of the elbow. JBJS. 2002;84(4):686–7.

38. Conwell H. Injuries to the wrist. Summit, NJ: CIBA Pharmaceutical; 1970.
39. Berger R. Anatomy and kinesiology of the wrist. In: Mackin E, Callahan AD, Skirven T, et al., editors. Rehabilitation of the hand and upper extremity. 5th ed. St. Louis: Mosby year book; 2002.
40. Kobayashi M, Berger R, Linscheid R, et al. Intercarpal kinematics during wrist motion. Hand Clin. 1997;13:143–9.

Kinetics and Kinematics of Rib Cage

9

9.1 Introduction

The biomechanics of rib cage is a very important segment as any alteration in the mechanics could lead to associated ventilatory dysfunction within the chest wall. In order to maintain proper ventilation, the mobility/kinematics of the rib cage is most important along with muscles that are responsible for driving ventilation. In this chapter, we shall learn about the important aspects of rib cage kinematics and kinetics along with few important pathomechanics.

9.1.1 Anatomical Background

The rib cage is a closed bony structure. The biomechanics of rib cage is well determined by the anatomical alignment of the associated structures within the thorax. The term "**costo**" describes the rib in anatomical studies. The most important anatomical landmarks that form the rib cage are the **sternum** (forms anterior border), **ribs** (forms lateral border), **thoracic vertebra** (forms posterior border), **jugular notch/clavicular notch and its attachment to costocartilage and the first rib** (forms the superior border), and **xiphoid process** (forms inferior border) [1].

It is also important to know the various types of ribs in the human body since they distinctly determine the biomechanical function. The rib cage consists of a total of 12 pairs of ribs. Ribs that attach directly to the sternum and the respective costocartilage are known as the true ribs (**first to seventh rib**) or **vertebrosternal** (attached to vertebra posterior and sternum anterior), whereas the ribs that attach indirectly to sternum are known as the **vertebrochondral** (attached to vertebra posterior and costocartilage of superior rib anterior) or **false rib (eighth, ninth, and tenth rib).**The rest two ribs **(11th and 12th)** are free anteriorly and do not attach to the sternum directly or indirectly and therefore are known as the **floating ribs**. The rib cage can be seen in C shape unilaterally since they gradually increase in size

A. Hazari et al., *Conceptual Biomechanics and Kinesiology*,
https://doi.org/10.1007/978-981-16-4991-2_9

from the first to seventh rib and then decrease from eighth to twelfth. Thus, it can be seen that the diameter of the rib is maximum at the midpoint of the rib cage in both sagittal and frontal planes. These anatomical positions determine normal function of breathing; any alteration would cause ventilatory disorders.

9.1.2 Rib Cage Articulations

The rib cage is attached to various structures forming the following **seven** joints [1] (Fig. 9.1):

(a) **Manubriosternal (MS)**
(b) **Xiphisternal (XS)**
(c) **Costovertebral (CV)**
(d) **Costotransverse (CT)**
(e) **Costochondral (CC)**
(f) **Chondrosternal (CS)**
(g) **Interchondral (IC)**

(a) **Manubriosternal joint:** It is a synchondrosis joint attached between manubrium and sternum [1].
(b) **Xiphisternal joint:** It is a synchrondosis joint attached between the xiphoid process and sternum [1].
(c) **Costovertebral joint**: It is a synovial joint attached between head of the rib and verterbral bodies [1]. The superior and inferior costovertebral joints are present in typical ribs (second to ninth ribs) as they articulate between the two adjacent vertebras (superior and inferior facets). Atypical ribs like the first and 10th-12th rib articulate with only one corresponding vertebral body [1, 2]. The rotation and gliding movements are seen in all costovertebral joints; however, atypical ribs allow more mobility than typical ribs [1, 3]. The CV joint has mobility and stability support from distinct ligaments such as **interosseous** (absent in atypical ribs) **and radiate ligaments**.

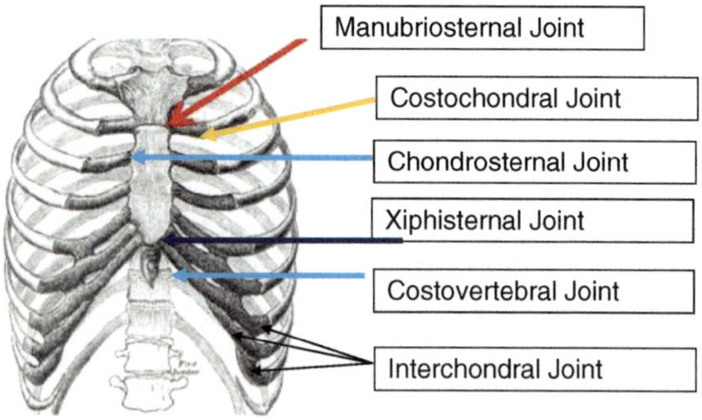

Fig. 9.1 Joints and articulations of the rib cage

(d) **Costotransverse joint:** It is a synovial joint attached between rib and transverse process of the vertebra. The CT joints are present from the first to tenth rib. Both rotation and gliding are seen at the CT joint. However, the rotation is more from first to seventh in contrast to gliding motion which is predominantly seen at seventh to tenth. Similar to CV joint, the CT joint capsule is reinforced by ligaments like **costotransverse ligaments** [1–3].
(e) **Costochondral joint:** It is a synchondrosis joint attached between rib and the corresponding costocartilage [1]. These joints are present anteriorly from the first to tenth rib. The CC joint lacks support from ligaments.
(f) **Chondrosternal joint:** It is a combination of synovial (second to fifth) and synchondrosis type (first, sixth, and seventh) attached between costocartilage and sternum [1, 2]. The important ligaments include **anterior** and **posterior radiate the costosternal ligament, sternocostal ligament,** and **costoxiphoid ligament** [1, 4].
(g) ***Interchondral Joints:*** *It is a synovial joint* attached between the two costocartilages and indirectly to the sternum (eighth to tenth ribs) [1].

9.2 Kinematics of the Rib Cage

We have learnt that the rib cage is attached to multiple joints, and thus, the kinematics of the rib cage depends upon significant contribution from each joint discussed above. There has been a conflict among researchers regarding the axis for rib kinematics. According to Levangie and Norkin [1], the upper ribs show sagittal plane movements predominantly as the axis is closer to the frontal. On the other hand, the lower rib movements are seen more in the frontal plane as the axis passes close to the sagittal.

9.2.1 Pump Handle Motion

The kinematics of the rib cage and thorax is significantly elicited during the breathing phases. The movement of ribs is seen more at joints on the anterior aspects compared to the joints that are present on the posterior aspects. As we have learnt above that the axis of movement for upper ribs is closer to the frontal direction, inspiration leads to elevation of the ribs in the sagittal plane. With the elevation of the ribs, the body of sternum is pushed forward and superior. As inspiration continues with maximum sternal excursion, the **anteroposterior diameter** of the thorax is increased. Since this motion is biomechanically similar to what is seen during a pump in borewell, the movement is known as the **pump handle motion** (Fig. 9.2). On the contrary, the lower ribs are attached more obliquely, and since the axis is more close to the sagittal direction, the motion of lower ribs would predominantly seen as the lateral elevation in the frontal plane increasing the **transverse diameter** of the thorax. This motion is similar to what

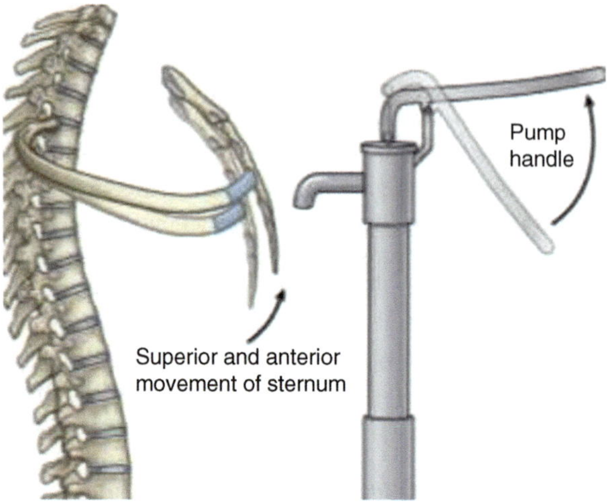

Fig. 9.2 Depicts pump handle motion at upper ribs (kinematics of the chest wall)

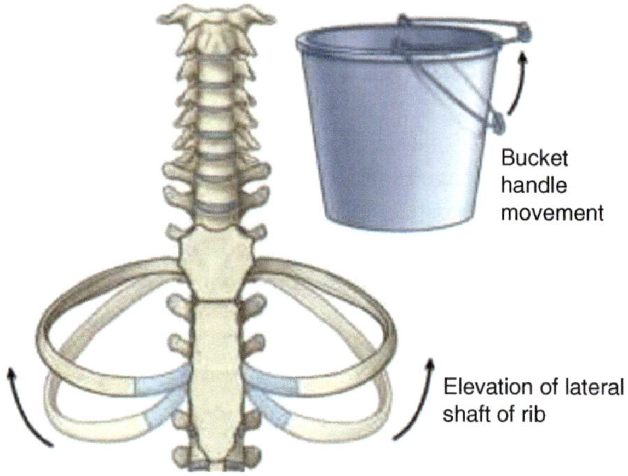

Fig. 9.3 Depicts bucket handle motion at lower ribs (kinematics of the chest wall)

is seen when a bucket handle is moved, and therefore, the lower ribs kinematics is also known as the **bucket handle motion** (Fig. 9.3).

It should be noted that the intermediate ribs show both pump handle and bucket handle movements due to change in the instantaneous axis during dynamic motions, whereas floating ribs do not participate in these motions [1].

9.3 Kinetics at Rib Cage

The muscles of the thorax are involved in ventilation and therefore are also known as the **ventilatory muscles**. The force generated by these muscles allows the pump and bucket handle motion under optimal expansion of the chest wall during inspiration. Since the expiration is passive, muscles are electrically silent in this phase. We shall learn about important muscles that act on the rib cage and how they function biomechanically. The unique features for the muscle of the rib cage include [1]:

1. **High oxidative content** (type I). Thus, it can work continuously without fatigue, e.g., diaphragm muscles keep contracting throughout life.
2. **Contraction is rhythmic** in nature and not episodic—contraction is based on the normal rhythm of inspiration and expiration rather than time bound. It is the respiratory drive that regulates the rhythmic contraction.
3. **Contraction against airway resistance** and the elastic properties of the lung is more compared to the gravitational force alone—muscles such as diaphragm and intercostals work on airway pressure gradient in addition to gravitational force where applicable.
4. **Ability to contract both voluntarily and involuntarily**—though the rhythmic contraction is involuntarily, we can also regulate the muscle voluntarily. For example, if we feel to increase airway we can voluntarily do a deeper and longer inspiration. This is known as **forced inspiration** and requires **accessory muscles** to take part in addition to **primary inspiratory** muscles.

9.3.1 Primary Muscle of Inspiration

Primary muscles of inspiration include **diaphragm, intercostals, and scalene** [1, 5].

(a) **Diaphragm:** it is the most important muscle of inspiration contributing to 80% of normal breathing [1, 6]. The biomechanical function of the diaphragm depends upon its orientation and attachment. The fibers originate from **(a) sternum, (b) costocartilage, (c) ribs, and (d) vertebral bodies**. Thus, we can see that the fibers arrangement would give an umbrella-like shape and when the umbrella is open completely the fibers get stretched and shift down maximum till an external resistance is applied to stabilize and stop further movement. This forms the basis for kinetics of the thorax as explained here. All the originating fibers of the diaphragm insert at center known as the central tendon which forms the **dome (top)**. The fibers originating from the sternum, ribs, and costocartilage are known as the **costal portion** of the diaphragm and attach to inner portion surface of lower six ribs. These fibers start from the lower ribs and ascend vertically up as **zone of apposition** and then become horizontal to insert to the central tendon forming dome [1, 7]. On the other hand, the fibers arising from the vertebral body and aponeurotic arcuate ligament are named as the **crural portion** (Fig. 9.4) [1, 8, 9].

Mechanism of action: During quiet breathing and normal inspiration, the diaphragm tends to increase the volume of lungs for inflow of air through the pressure gradient (if space inside the thoracic cage increases it would create a negative pressure for airflow from outside to inside). In the event the zone of apposition contracts and descends till the point the dome compresses the abdominal contents, it leads to an increase in intra-abdominal pressure. With maximum descent

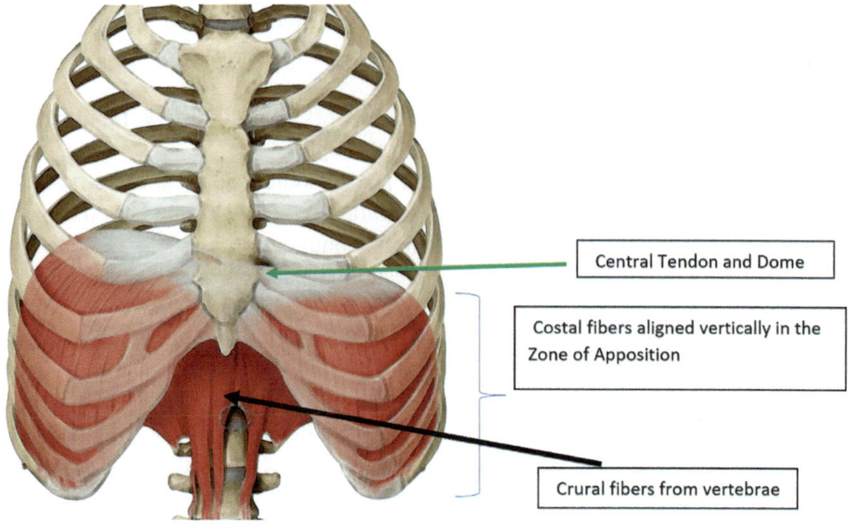

Fig. 9.4 Parts of the diaphragm and its anatomical connections

the dome stabilizes over the abdominal contents; therefore, no further downward movement of the muscle can take place. However, the continued contraction of the costal fibers would now move the origin end instead of the insertion, where the lower ribs would be lifted and rotated outward leading to the bucket handle movement [1, 10]. Further contraction would not be possible as the outward lifting of ribs would stretch the costal fibers maximally (horizontal) at both insertion and origin [11]. The crural portion is less active compared to costal and helps increasing intra-abdominal pressure while contracting to descend the central tendon indirectly [1, 9]. A sequence of events for the role of diaphragm in thorax kinematics has been given in flowchart (Fig. 9.5).

(b) **Intercostal muscles:** The parasternal muscle is considered as the primary muscle for inspiration. The internal, external, and lateral intercostal muscles are part of intercostals which are often referred to as ventilator muscles of respiration. Both internal and external intercostal muscles have been shown to be active in EMG studies during inspiration and expiration [1, 12]. The proposed action of the external intercostal muscle is to lift the lower rib and that of internal intercostals is to lower the higher ribs to increase the chest volume. The pattern of recruitment is from higher intercostal muscle to the lower intercostals [1, 13].

(c) **Scalene muscles:** As the primary muscle of inspiration, the scalene action is to lift the sternum and first two ribs in the pump handle motion [14]. The scalene can generate more force at end inspiration when diaphragm force decreases. This is the reason we observe the lateral expansion of the chest during the initial phase of inspiration followed by anteroposterior expansion at the end of inspiration. The scalene also works in association with parasternal to counteract

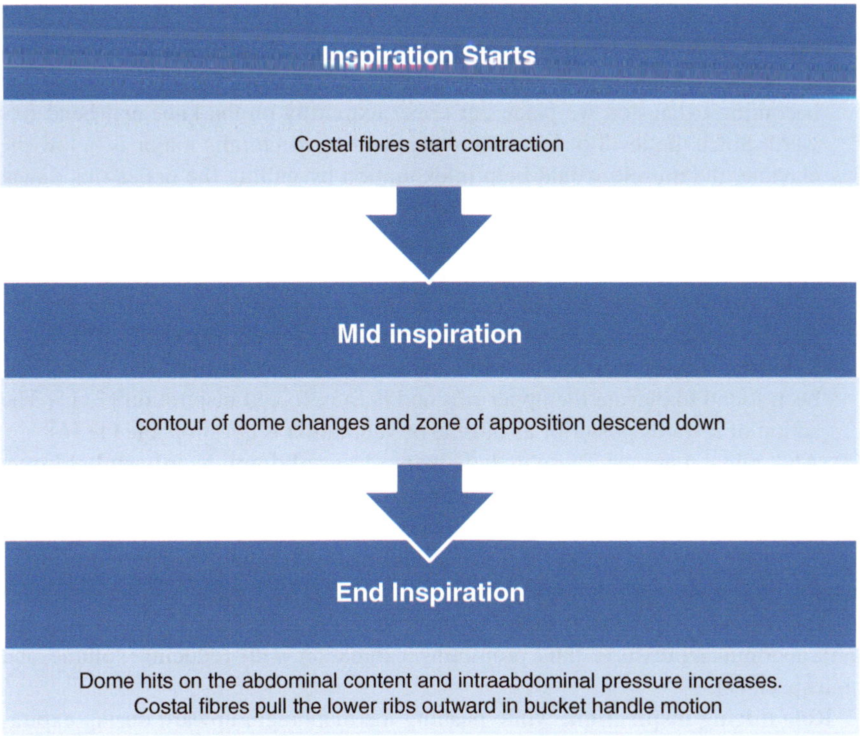

Fig. 9.5 Sequence of events for diaphragm action during inspiration

the paradoxical movement of the upper chest against diaphragm contraction and decreased intrapulmonary pressure.

9.3.2 Accessory Muscles

Accessory muscles include **sternocleidomastoid, upper trapezius, pectoralis major and minor, subclavius, levatores costarum, and abdominal muscles**. The accessory muscles of inspiration help to meet the increased demand of oxygen in need. The muscles attach the rib cage to the shoulder, head, vertebral column, or pelvis such that when the trunk is stabilized the rib cage can be mobilized upward and outward to supply extra ventilation to lungs [1, 8].

(a) **Sternocleidomastoid and trapezius:** The attachment of the muscle to the neck and clavicle allows the upper ribs to be pulled up (pump handle motion) and increase the anteroposterior diameter of the thorax. During running and high-intensity exercise, we can see the strong contraction of sternocleidomastoid and trapezius as the muscles bulge out.
(b) **Pectoralis major and minor:** The accessory action of pectoralis major depends upon the type of fiber. The **sternal fibers** pull the rib cage up when upper

extremity is fixed as seen during exercises like pull-ups. The role of **clavicular fibers** can be inspiratory or expiratory depending upon its attachment level on the humerus with respect to clavicle. For example, after finishing running and becoming exhausted we place our upper extremity on the knee and bend forward. Since the level of humeral attachment for pectoralis major is below the clavicle, the muscle would help in expiration by pulling the upper ribs down. On the contrary, when we sprint, the additional movement of reciprocal hand with elbow flexed allows pectoralis major to act as an accessory inspiratory muscle while pulling upper ribs and clavicle up to meet the increased demand of ventilation. In addition, the pectoralis minor muscle helps to pull the third to fifth rib upward along with the subclavius during forced inspiration ([1].

(c) **Levatores costarum** and serratus posterior**:** Levatores costarum action has been found to elevate the upper ribs and help in forced inspiration [1, 15] The action of serratus posterior as muscle of ventilation is questionable [1, 16].

(d) Abdominal muscles: These include **transversus abdominis, internal oblique, external oblique, and rectus abdominis**.

The major function of the abdominal muscles with regard to ventilation is to assist with forced expiration and phased inspiration [1].

Role during expiration: helps to push diaphragm upward by increasing the intra-abdominal pressure, thus promoting exhalation with reducing volume and high speed [1].

Role during inspiration: while pushing the diaphragm upward during expiration, there is a passive stretch on the costal fibers which drives the next inspiration so that the length–tension relationship of the diaphragm is maintained. Also the descent of the dome of the diaphragm during inspiration is stabilized by contraction of abdominal muscles while maintaining **compliance** [1].

9.4 Pathomechanics

The biomechanics of the chest wall is susceptible to various pathological conditions. The structural changes in the rib cage can also lead to altered kinematics and kinetics. We shall discuss few important conditions here.

Altered compliance: The compliance of an organ is the measure of its distensibility. During inspiration the abdominal volume and pressure change which determines its distensibilty or compliance. If we look at the rib cage as volumetric closed chamber system, the compliance of the abdomen is the ratio change in volume to change in pressure. Now let us understand two clinical conditions where the altered compliance could affect the biomechanical function of the rib cage.

1. **Spinal cord injury:** In spinal cord injuries below C3, when phrenic nerve innervations are lost and thus the diaphragm also cannot contract, the compliance of the chest wall increases as the volume is high. Also, the intra-abdominal pressure is not maintained by contraction of the diaphragm muscle, and therefore, the

central tendon cannot stabilize. As a result, the lateral expansion/bucket handle movement of lower ribs will be affected.
2. **Pregnancy:** During pregnancy the diaphragm is pushed into a higher position as the abdominal contents distend. With inspiration the dome and central tendon cannot descend much and get stabilized over the abdominal contents at a relatively higher position than normal which would cause unwanted lateral expansion of ribs in the early inspiratory phase and affect the tidal volume.
3. **Chronic obstructive pulmonary disease (COPD):** The major respiratory dysfunction among patients with COPD is loss of lung elasticity. As we have learnt that expiratory is passive due to elastic recoil of the lungs which pushes the air out, the loss of such functions as seen in COPD will lead to extra volume of air inside and cause lung hyperinflation. In addition, the diaphragm attains a lower position in COPD. As a result of hyperinflation, the anteroposterior diameter of the chest increases and a barrel chest is seen. Also, the persistent downward position of the diaphragm weakens the costal portion and thereby the zone of apposition becomes completely horizontal. Now with increased efforts to contract the diaphragm during inspiration, the costal portion would pull the ribs inward than doing its normal function of lateral elevation.

9.5 Morphological Difference for Elderly and Infants

The structure and function of the chest wall and ribs differ a lot at different stages of life which determines its kinematics and kinetics. For instance, the distensibility of an infant's chest wall is more so that they can adapt to structural changes during birth. With growing age, the chest wall becomes stiffer and distensibility reduces as seen in the elderly. Let us look at the comparative difference for an infant and an elderly compared to a young adult (Table 9.1).

Table 9.1 Comparison of the chest wall for infants and the elderly to healthy adults

Chest wall features in infants	Chest wall features in the elderly
Higher compliance	Lower compliance
Stabilizers over mobilizers	Hypermobility and susceptible to more fractures
Delayed ossification of ribs compared to other bones at respective age	Fibrosis of joints
More horizontal alignment of ribs compared to adults	Loss of elastic recoil increases anteroposterior diameter of the chest
Diaphragm is attached horizontally compared to elliptical in adults	Loss of muscle strength affecting lung volume and capacity
Lower ribs gets pulled inward than outward making the chest distorted during inspiration	Diaphragm is positioned lower and the number of type II fibers reduces
The range of motion for chest wall expansion is less	Over and early recruitment of accessory muscles
Lower number of type I muscle fibers in diaphragm making it fatigued easily on stress	Abdominal compliance decreases to produce effective intra-abdominal pressure
Accessory muscles are inefficient	Range of motion for the chest wall decreases

9.6 Summary

The kinematics and kinetics of the chest wall is important to understand the ventilator function. The structure of the ribs and their articulations allow the chest wall to expand during inspiration. The muscles of the chest wall are actively involved in the kinetic function. Biomechanical changes in the structure and function of the chest wall could interfere with respiratory function of the human body.

References

1. Levangie PK, Norkin CC. Joint structure and function: a comprehensive analysis. Philadelphia, PA: F.A. Davis; 2011.
2. Winkel D. Diagnosis and treatment of the spine. Gaithersburg, MD: Aspen; 1996. p. 393–401.
3. Kapandji IA. The physiology of the joints: the trunk and vertebral column. 3rd ed. New York: Churchill Livingstone; 1990.
4. Grieve GP. Common vertebral joint problems. 2nd ed, pp 32–39, 110–129. New York: Churchill Livingstone; 1988.
5. Tobin MI. Respiratory muscles in disease. Clin Chest Med. 1988;9:263–86.
6. De Troyer A, Estenne M. Coordination between rib cage muscles and diaphragm during quiet breathing in humans. J Appl Physiol. 1984;57:899–906.
7. De Troyer A, Estenne M. Functional anatomy of the respiratory muscles. Clin Chest Med. 1988;9:175–93.
8. Celli BR. Clinical and physiologic evaluation of respiratory muscle function. Clin Chest Med. 1989;10:199–214.
9. De Troyer A, Sampson M, Sigrist S, et al. The diaphragm: two muscles. Science. 1981;213:237–8.
10. Epstein S. An overview of respiratory muscle function. Clin Chest Med. 1994;15:619–38.
11. Reid WD, Dechman G. Considerations when testing and training the respiratory muscles. Phys Ther. 1995;75:971–82.
12. LeBars P, Duron B. 1984: Are the external and internal intercostal muscles synergistic or antagonistic in the cat? Neurosci Lett. 1984;51:383–6.
13. Koepke GH, Smith EM, Murphy AJ, et al. Sequence of action of the diaphragm and intercostals muscles during respiration. I. Inspiration. Arch Phys Med Rehabil. 1958;39:426–30.
14. De Troyer A. Actions of the respiratory muscles or how the chest wall moves in upright man. Bull Eur Physiopathol Respir. 1984;20:409–13.
15. Goldman MD, Loh L, Sears TA. The respiratory activity of human levator costal muscle and its modification by posture. J Physiol. 1985;362:189–204.
16. Vilensky J, Baltes M, Weikel L, et al. Serratus posterior muscles: anatomy, clinical relevance and function. Clin Anat. 2001;14:237–41.

Kinetics and Kinematics of Vertebral Column

10

10.1 Introduction

The vertebral column of the human body consists of distinct segments like cervical, thoracic, lumbar, sacral, and coccyx vertebra. There are 33 vertebrae constituted by seven in the cervical, twelve in the thoracic, five in the lumbar and five in the sacral, and four in the coccygeal segments, respectively. Biomechanically the vertebral column is the most important segment because it is the link between the upper extremity and the lower extremity where stability and mobility remains constantly challenged. The correct biomechanics is thus important to transfer the load to the lower segments. In this chapter, we shall focus on the kinematics and kinetics of the vertebral column.

10.1.1 Anatomical Background

The vertebral column has two major shapes. The **primary curve** also known as the "**kyphotic curve"** is convex posteriorly like a mirror image of the letter C and it is present from birth. The kyphotic curve remains constant at the thoracic and sacral region of the vertebral column. The **secondary curve, also** known as the "**lordotic curve,"** is convex anteriorly and develops at cervical and lumbar segments of the vertebral column in the later stage of life. These curves are not present at birth and develop over time when we start to stand upright against gravity from the crawling position.

10.1.1.1 Articulations of the Vertebral Column in General

There are two major articulations present at the vertebral column. The articulation between two vertebral bodies is known as the **intervertebral/interbody joints**. They are cartilaginous joint [1]. The articulation between the facets is known as the **zygapophyseal joints** which are synovial.

A. Hazari et al., *Conceptual Biomechanics and Kinesiology*, https://doi.org/10.1007/978-981-16-4991-2_10

10.1.1.2 Ligaments of the Vertebral Column in General

In general, there are six main ligaments of the vertebral column which play an important role in the kinematics and kinetics of the intervertebral and zygapophyseal joints. These include **anterior longitudinal ligament (ALL), posterior longitudinal ligament (PLL), ligamentum flavum, supraspinous ligament, interspinous ligament, and intertransverse ligament** [1].

10.1.2 Role of Intervertebral Disks

In order to distribute the load and allow smooth mobility, 23 intervertebral disks are present between the vertebrae. The mobility function of the disk is determined by its size. In addition, the size of the disk reflects the load-bearing capacity. For instance, the size is less in the cervical region compared to the lumbar where higher load bearing is required [2]. The size of the disk also adds to the overall length of the vertebral column. However, the available range of motion at the vertebra is determined by the ratio of disk thickness and vertebral body height which is highest in the cervical region, and thus, the range of motion is also greater compared to the other segments of the vertebral column [2]. Mobility at the thoracic region is least since it has stability function for the rib cage biomechanics. The load-bearing function of the disk is taken care by its ability to deform. The disk has three major components like the **nucleus pulposus, annulus fibrosus, and end plate** (Fig. 10.1a). The nucleus pulposus is the central gelatinous substance that can easily change its shape when loaded. The annulus fibrosus is fibrous ring outside the nucleus pulposus and helps to resist the tensile forces. The vertebral end plate is the outermost part that is responsible for resisting compressive forces. The kinetic function of the respective parts of disk is determined by the composition and type of ground substance present in them which has been explained in the previous chapters. We shall learn the mechanics of load-bearing function here. When the load is applied from the vertebral bodies to the disk, the nucleus pulposus reacts by deforming its shape and bulging out like balloons and putting pressure on the annulus fibrosus. The annulus fibrosus in turn puts an equal force toward the nucleus to resist the tensile force and maintain the equilibrium. The force when exceeds beyond the annulus strength is then transmitted to the end plate [1].

10.2 Kinematics of the Vertebral Column in General

The vertebral column has the ability to move through all axis and all planes, thus showing six degrees of freedom. The predominant motions include **flexion, extension, side flexion to left, side flexion to right, rotation to left, and rotation to right** (Fig. 10.1b). It should be noted that the motion occurring in the vertebral column is not a pure single planar motion but rather a coupled motion. For example, lateral flexion is always coupled with rotation [1]. **Coupling** is defined as the consistent association of one motion about an axis with another motion around a different axis [1].

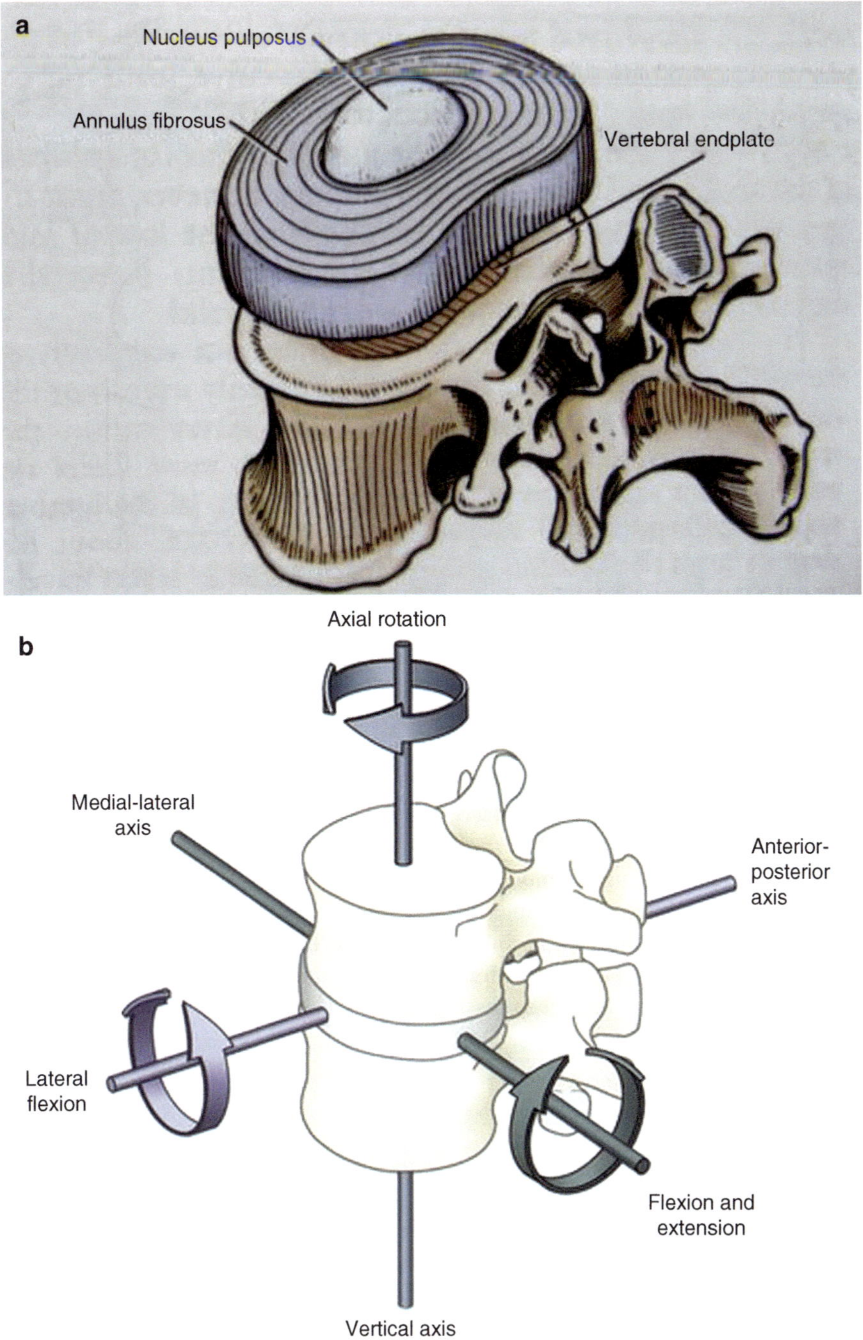

Fig. 10.1 (**a**) Parts of intervertebral disk. (**b**) Motions of the vertebral column

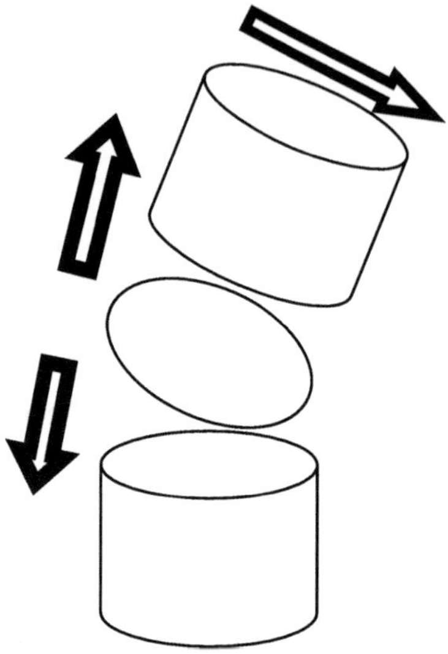

Fig. 10.2 Showing flexion at the vertebral column (anterior tilting and gliding of superior vertebra)

(a) **Flexion:**

Osteokinematics: The motion occurs in the sagittal plane and frontal axis. There is separation of spinous process at the adjacent vertebrae.

Arthrokinematics: The flexion at the vertebral column is seen as anterior tilting and gliding of the superior vertebra over the inferior [1]. The intervertebral foramen increases with flexion. Therefore, flexion exercises are indicated in cases of stenosis or any condition that reduces the intervertebral foramen size (Fig. 10.2).

Facilitators: Intervertebral disk (the size of the disk in the individual region would determine the range of motion).

Limiters: Supraspinous and interspinous ligaments resist separation of spinous process, thereby limiting flexion. In addition, passive structures which are located posterior such as the joint capsule, ligamentum flavum, PLL, and posterior part of annulus fibrosus would restrict excessive flexion [1]. Active tension is created by back muscles of the trunk.

(b) **Extension:**

Osteokinematics: The motion occurs in the sagittal plane and frontal axis where spinous process comes closer.

Arthrokinematics: The extension at the vertebral column is seen as posterior tilting and gliding of the superior vertebra over the inferior (Fig. 10.3). The size of the intervertebral foramen decreases. Therefore, extension exercise is indicated in case of prolapsed disk.

Facilitators: Intervertebral disk (the size of the disk in the individual region would determine the range of motion).

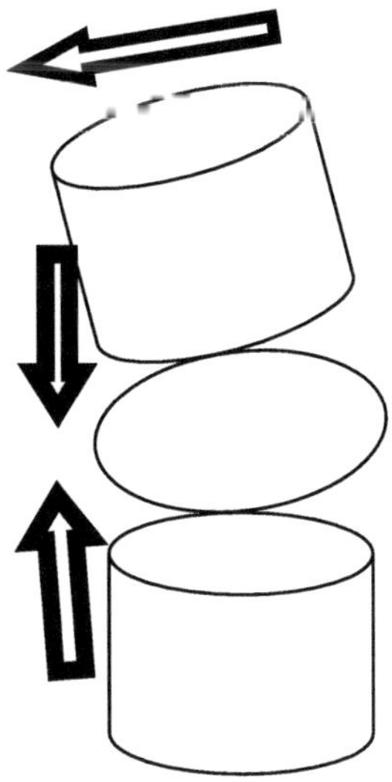

Fig. 10.3 Showing extension at the vertebral column (posterior tilting and gliding of superior vertebra)

Limiters: Passive tension is created by bony contact of spinous process, joint capsule, ALL, and anterior part of the annulus fibrosus [1]. Active tension is created by anterior muscles of the trunk and abdomen.

(c) **Side flexion/lateral flexion:** As discussed above, the lateral flexion of the spine is a coupled motion; both gliding and rotation are observed during motion.

Osteokinematics: The motion occurs in the frontal plane and sagittal axis. The transverse process toward the side of flexion gets compressed and the opposite side gets stretched. In addition, the annulus fibrosus on the side of concavity gets compressed.

Arthrokinematics: During lateral movements, flexion and rotation of the superior vertebra take place over the inferior one. For example, if we do lateral flexion to left, the superior vertebra will flex and rotate toward left (ipsilateral).

Facilitators: Intervertebral disk (the size of the disk in the individual region would determine the range of motion). There is active contraction of lateral trunk muscles on the side of lateral flexion.

Limiters: Passive tension of the annulus fibrosus on the contralateral side; intertransverse ligament limits the lateral flexion movement at spine. In addition, the tension in trunk muscles on the opposite side would act as the limiters as well.

(d) **Side Rotation:** The rotation movement in the spine is also a coupled motion where rotation and gliding take place simultaneously.

Osteokinematics: The motion occurs in the transverse plane and vertical axis. The structures toward the side of rotation get relaxed and the opposite side gets stretched.

Arthrokinematics: During lateral rotation, flexion and rotation of the superior vertebra take place over the inferior one as coupled motion. For example, if we do lateral rotation to left, the superior vertebra will rotate toward left (ipsilateral) and glide into lateral flexion toward the right (contralateral) [1, 3].

Note: It should be understood that motion and its range on the vertebral column would also depend upon the segmental variation of the vertebral segments. Though the general kinematics remains the same, for reference to anatomical differences in vertebrae at cervical, thoracic, lumbar, and sacral segment, please refer any textbook of anatomy. Here, we shall concentrate on the biomechanical features directly.

10.3 Kinetics of the Vertebral Column in General

The vertebral column undergoes different types of forces. The main kinetic function of the vertebral column is to distribute the load efficiently. Studies have shown that the vertebral column is exposed to **axial compression, torsional force** (during coupled motion)**, shear forces** (during flexion/extension and gliding motions)**, tensile force** (lateral flexion), etc., in both dynamic and static state. Thus, the study of spine kinetics is an important aspect to understand the pathomechanics and design their rehabilitation strategies. Let us understand the response of the vertebral column under specific forces like compressive and shear.

(a) **Compressive force:** The axial compression on the spine is the most evident by the body weight and force of gravitation. The force acts through the long axis of the column from top to bottom. The anatomical flexible curvature has an advantage to tackle this force in comparison to a rigid rod-like structure. The major resistance to axial compression is taken by the disk and the vertebral bodies. As we have discussed before, the vertical loading causes a deformation in the disk. As the pressure increases, the nucleus pulposus bulges out creating a swelling pressure which is transmitted to the nucleus fibrosus. The stress created at the annulus fibrosus with pressure from the nucleus is controlled by the resistive tensile strength of the annulus and very little pressure is transmitted to the vertebral body through end plates. This is the reason that disk is susceptive to prolapse with heavy weight lifting and altered biomechanics. Under normal stress, the disk and vertebral bodies can sustain axial loading very efficiently.

    The vertebral bodies of the lumbar spine are bigger and therefore can take more axial loading in a young adult. The zygapophyseal joint and spinous process also act as accessory structure to withstand compressive loading [1]. With aging over 40 years, compressive strength and stiffness decrease [1, 4].

    Clinical application: The vertebral column length shows diurnal variation. The reason for the same is attributed to the viscoelastic property, i.e., creep of the disk. In instances where the axial loading takes place over an extended period of time (standing for long), the nucleus pulposus exhibits high swelling pressure. Thus, the fluid escapes into the annulus part which returns back on rest. As a result, an individual height may differ when he gets up in the morning after rest (maximum height of disk) and during the day (height loss of disk). An optimal creep is required for healthy nutrition of the disk; however; excessive and prolonged creep will reduce the compressive stiffness and make the disk susceptible to bulge or mechanical injury.

(b) **Shear force:** We have learnt that shear forces are linear force system in the direction of movement. At the vertebral column, the shear force is seen during the translator or gliding movement predominately occurring at flexion and extension. The shear force at the spine is controlled by the facet joint and disk.

(c) **Bending force:** The bending of the spine causes compression on the side of motion and tensile force on the opposite side. For example, during flexion, the anterior structures get compressed and posterior structures get tensed. The anterior and posterior longitudinal ligaments provide stability to bending forces.

10.4 Kinematics and Kinetics of the Regional Segments

10.4.1 Cervical Spine

The cervical spine has seven vertebrae in total and divided into two segments: **upper cervical** (craniovertebral) and **lower cervical spine**.

10.4.1.1 Craniovertebral Articulation

It consists of occipital condyles and **C1 (atlas), C2 (axis) vertebra.** The upper cervical segments include two joints: atlanto-axial (between C1 and C2) and atlanto-occipital (between C1 and occiput). Only the nodding movement is available at atlanto-occipital joint, whereas atlanto-axial shows axial rotation predominantly.

Craniovertebral ligaments: The ligaments of the cervical spine are extension of the ligaments from the lumbar and thoracic spine.

(a) **Anterior atlanto-occipital and atlanto-axial membrane:** Extension of the anterior longitudinal ligament which resists excessive extension.

(b) **Posterior atlanto-occipital and atlanto-axial membrane:** Extension of the ligamentum flavum and helps increasing the range of cervical rotation significantly [1].

(c) **Tectorial membrane:** Extension of the posterior longitudinal ligament (PLL) which resists excessive flexion.

(d) **Ligamentum nuchae:** Extension of the supraspinous ligament which can resist lateral flexion.

(e) **Transverse ligament/ atlantal cruciform ligament:** Very strong ligament which is attached to dens and stabilizes the C1/C2 segment of the cervical spine.

(f) **Alar ligament:** Attaches occipital condyle and atlas. It restricts flexion, lateral flexion, and rotation of the atlanto-occipital joint [1, 5].

(g) **Apical ligament:** Connects axis to the occiput and forms the anterior margin of the foramen magnum [1, 6].

10.4.1.2 Lower Cervical Spine Articulation

It consists of C2 to C7 vertebrae with articulation between the superior and inferior facets known as the superior zygapophyseal and inferior zygapophyseal joints. Kinematics include flexion, extension, side flexion, and side rotation movements.

10.4.1.3 Kinetics at the Cervical Spine

The cervical segment is biomechanically designed for mobility. Therefore, the magnitude of axial loading (weight of skull), tension (ligaments and capsule), and shear stress is less compared to the lumbar region. The upper cervical vertebra does not have disk and the forces are directly transferred to the neural arch where the lamina takes the maximum compressive loading. The disk is present at the lower cervical segments and helps to distribute the axial loading as discussed above. In addition, the vertebral bodies and zygapophyseal facet help to distribute and counter the stress imposed on the cervical segment. Studies have shown variation in magnitude of compressive loading with respect to body posture. At standing and sitting, the loading is less compared to end range of flexion and extension [7]. This could be attributed to the angular gravitation forces and higher counter force of muscle, ligaments, and capsules to maintain the position. In comparison to the lumbar segment, the cervical segments show lesser stiffness for bending and torsion forces. However, the axial compressive stiffness is found to be similar [8]. This suggests that the head should not be rotated during activities of axial loading such as flexion and extension to minimize the risk of injury.

10.4.1.4 Kinematic Function of Muscles at Cervical Spine

The muscles of the cervical spine generate forces to move the joint as well as stabilize them. It should be noted that muscles having attachment toward the side of motion will act concentrically and muscles that attach opposite would act eccentrically. The important muscles of the cervical region have been listed here [1].

Flexors: Longus capitis, anterior scalene muscles, longissimus capitis, and longissimus cervicis.

Extensor: Suboccipital muscles and trapezius when upper extremity is fixed. Splenius capitis and splenius cervicis (bilateral contraction), semispinalis capitis, and semispinalis cervicis.

Lateral flexors: Posterior scalene muscles, levator scapulae (when the upper extremity is fixed and passively it resists the anterior shear force of gravity, suboccipital muscles (unilateral action), and anterior scalene muscles (ipsilateral flexion working unilaterally).

Rotators: Levator scapulae, splenius capitis and splenius cervicis (ipsilateral rotation during unilateral contraction), suboccipital muscles (ipsilateral rotation during unilateral contraction), and anterior scalene muscles (contralateral rotation working unilaterally).

10.4.2 Thoracic Spine

10.4.2.1 Articulation

Interbody Joints

The articulation between the vertebral bodies of the thoracic segments forms the respective interbody joints. The presence of intervertebral disk which is thinner in comparison to lumbar allows limited motion in the thoracic vertebra.

Zygapophyseal Joint

The articulation at the thoracic region is mainly seen at the superior and inferior zygapophyseal facets and termed as the superior and inferior zygapophyseal (facet) joints. The facet joints of the upper thoracic segments lie closer to the frontal plane, whereas the lower segments lie close to the sagittal plane determining the kinematics of the thoracic segments. Clearly, the upper thoracic segments would allow lateral flexion and rotation over the flexion and extension as they are closer to the frontal plane.

10.4.2.2 Ligaments at Thoracic Spine

They include anterior longitudinal ligament (ALL), posterior longitudinal ligament (PLL), ligamentum flavum, supraspinous ligament, interspinous ligament, and intertransverse ligament. The ligaments of the thoracic spine are common as seen at the lumbar and cervical segments. The ligamentum flavum and ALL have been found thicker in comparison to the cervical region [1].

10.4.2.3 Kinematics at Thoracic Spine

Although all motions (flexion, extension, lateral flexion, and side rotation) are available at the thoracic spine, the range of motion is significantly less as the segments are designed to provide more stability over mobility. The lower thoracic joint lies closer to the sagittal plane and thus shows greater flexion and extension range of motion compared to the upper joints. Coupled motion is seen for lateral flexion and rotation. However, the range of coupled rotation during lateral flexion decreases as we have learnt that the facet is close to the sagittal plane. In the upper thoracic segment, lateral flexion and rotation occur in the same direction in contrast to the lower segment where lateral flexion is contralateral to rotation [9]. The rotation motion of the thoracic region is unique and determined by the ability of the rib cage to change its shape (distortion). During motion, the posterior rib becomes convex on the side of rotation and anterior ribs become convex on the opposite side.

10.4.2.4 Kinetics at the Thoracic Spine

The thoracic spine has been found to withstand more compressive force in comparison to cervical due to increased segmental body weight as well as the line of gravity passing distant to center of rotation causing a greater external flexion moment. As a result the ligaments located posteriorly and muscles of extension are under higher stress to counter the external moment. We shall discuss the muscles of the thoracic spine under the lumbar section as they have common function.

10.4.3 Lumbar Spine

10.4.3.1 Articulation

Interbody Joints
Joints formed between the vertebral bodies.

Zygapophyseal Joints
The facet joint in the lumbar region is synovial and the orientation is biplanar. The anterior portion lies in the frontal plane and the posterior portion is closer to the sagittal plane. Thus, the lumbar region shows greater mobility in all directions compared to thoracic segments; however, the range of motion is less compared to the cervical segments.

Lumbosacral Joint
The articulation between the fifth lumbar vertebra and first sacral vertebra is known as the lumbosacral joint. The angle of the lumbosacral joint has biomechanical significance as it determines the amount of shear force at the joint. The increase in angle would not only increase the lumbar lordosis but also cause excessive anterior shearing which may lead to slipped disk. This is the reason disk prolapse is very common at this segment explained by the vector diagram in the pathomechanics section ahead.

10.4.3.2 Ligaments at Lumbar Spine
In addition to all the ligaments mentioned above, the lumbar segments have specific ligaments such as **iliolumbar ligament and thoracolumbar fascia.** The thoracolumbar fascia (also called the lumbodorsal fascia) consists of three layers: the posterior, middle, and anterior. The fascia is very important and serves to transmit the force between the pelvis and trunk [10].

10.4.3.3 Kinematics at Lumbar Spine
Flexion and extension: Flexion and extension motion at the lumbar spine are pure planar motions significantly seen due to more sagittal plane orientation of the facets. However, the flexion range of motion is more compared to extension and follows the same tilting and gliding arthrokinematics as explained in previous sections.

Rotation and lateral flexion: The superior facets in the lumbar region are oriented in the sagittal plane, thus restricting rotation. However, the lateral flexion is evident and coupled with rotation. For example, if we flex laterally at the lumbar spine, the vertebra would also rotate ipsilaterally. In contrast, the axial rotation is accompanied by the contralateral side flexion.

Lumbopelvic rhythm: The integrated motion of the lumbar spine with the pelvis to increase the range of motion is termed as the lumbopelvic rhythm. This can be understood with an example illustrated here (Fig. 10.4a). If we try to bend forward from the hip joint completely and keep the lumbar spine straight, the range of motion is less and we cannot touch the ground. If we allow the lumbar segment to flex beyond the pelvis range of motion, we shall be able to touch the ground and

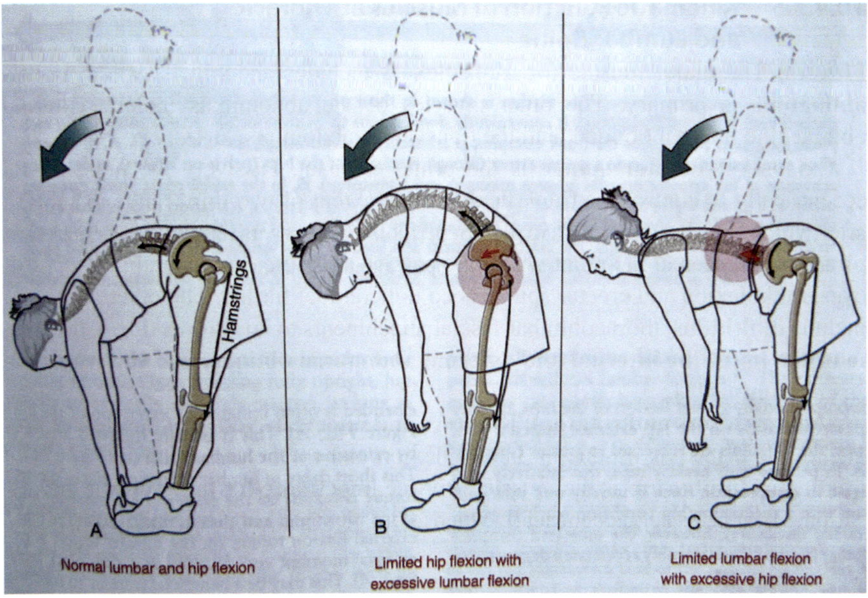

Fig. 10.4 Depicts lumbopelvic rhythm [11]

increase our range of motion. This is known as the lumbopelvic rhythm. It is important to understand that both segments need to contribute to the range of motion. The limited range of motion of either segment would not allow the lumbopelvic rhythm to take place smoothly (Fig. 10.4b and c) [12].

10.4.3.4 Kinetics at the Lumbar Spine

Axial loading: The lumbar spine is subjected to maximum axial loading and compressive force in comparison to the cervical and thoracic segments. The vertebral bodies and the disk play the most important role to withstand this stress. We have learnt that height of the vertebral body as well as the height of the disk at the lumbar region is maximum compared to the cervical and thoracic spine which helps to redistribute the force evenly. Studies have shown that the interbody joints and disk can withstand 80% of the compressive load, whereas the facet joints take up to 20% [1, 13]. The normal lordotic curve is important to maintain the compressive stress on the lumbar spine. It has been found that increased lordosis leads to increased compressive stress caused by the ground reaction force, muscular forces of the erector spinae, and abdominals [1, 14]. Thus, we find that the muscles of the lumbar segments are also actively involved in producing the stress through stronger contraction.

Shear force: The lumbar segments are subjected to constant shearing force during the action of the ground reaction force and lordotic curve. The shearing force is restricted by the bony contact of the facet joint under normal biomechanics. Any alteration in the lumbar curve would cause the shearing force to rise and failure of the joint structure and function.

10.4.3.5 Kinematic Function of Muscles at Thoracic and Lumbar Spine

Flexors: The major muscle for trunk flexors includes abdominals where **rectus abdominis** is primary. The other muscles of the abdominals include external oblique, the internal oblique, and the transversus abdominis.

Extensors: Erector spinae is the prime extensor. Passively the muscle acts eccentrically to control the trunk flexion to an extent of two-third range of motion after which they become electrically silent. This is known as the **flexion-relaxation phenomena**, seen at the point where the passive structure can maintain the control over trunk flexion and erector spinae need not contract further. The other extensors include **multifidus,** thoracolumbar fascial attachments to **latissimus dorsi, the gluteus maximus, the internal and external abdominal oblique, and the transversus abdominis** [1].

Lateral flexors and rotators: Rotators and intertransversarii muscles are frequently described as producing lateral flexion and rotation, respectively [1]. The **quadratus lumborum** is the muscle of prime importance for frontal plane stabilization. The muscle acts eccentrically when lateral flexion to opposite side takes place.

10.4.4 Sacroiliac Joint

10.4.4.1 Articulations

Sacroiliac Articulations

The sacroiliac joint (SI joint) is the articulation between the fused sacral vertebra and iliac bone bilaterally (forming left and right SI joint). The articular surface on the sacral and iliac crest is C-shaped (Fig. 10.5).

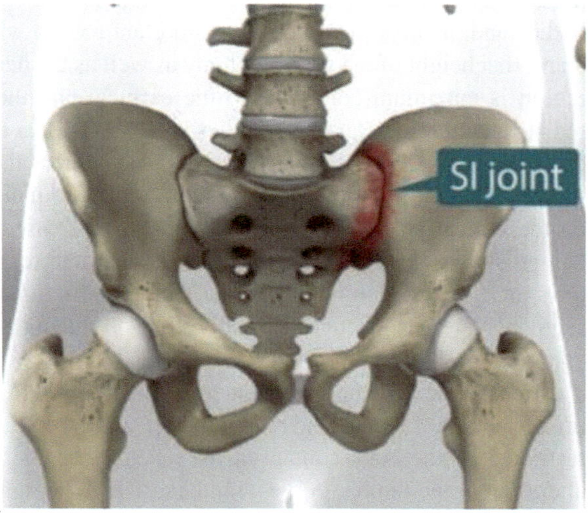

Fig. 10.5 Showing sacroiliac joint

Symphysis Pubis Articulation

The symphysis pubis is a cartilaginous joint located between the two ends of the pubic bones [1].

10.4.4.2 Ligaments of SI Joints

The ligament of the SI joint includes **sacroiliac ligament, iliolumbar ligaments, sacrospinous ligaments,** and the **sacrotuberous ligaments**. The ligaments are responsible for providing stability to the SI joint predominantly.

10.4.4.3 Ligaments of Symphysis Pubis Joints

Superior pubic ligament, inferior pubic ligament, and posterior ligament.

10.5 Kinematics of SI Joint

The sacroiliac joint allows minimal translation and gliding moments in the sagittal plane termed as the **nutation and counternutation** motions.

10.5.1 Nutation

The motion is seen as the anterior and inferior glide of the sacrum such that the coccyx tips posteriorly. The nutation motion increases the anteroposterior diameter of the pelvic outlet [1].

This is important during pregnancy where the motion of the SI joint should be more and adequate space can be provided to the growing fetus.

10.5.2 Counternutation

The counternutation motion is in the opposite direction of nutation where the coccyx tilts anteriorly and the sacrum moves posterior. Thus, the anteroposterior diameter of the pelvic brim is increased.

10.5.3 Clinical Implication of Nutation and Counternutation

The nutation and counternutation motions are important biomechanical function used during labor. The application of nutation and counternutation helps clinicians to process a smooth normal delivery. In the initial stage (motion of hip over pelvis), extension of the hip causes counternutation and increases in the pelvic brim which is required for the fetus to descend during labor. At a later stage, hip is flexed for inducing nutation movement with an increase in pelvic outlet and easy passing of the baby [1, 15].

10.5.4 Kinetics at SI Joint

The SI joint functions to bear a significant amount of body weight and therefore undergo compressive stress. However, the line of gravity and body weight creates a torsion stress which is stabilized by the surrounding ligaments and muscles. It has been proposed that the line of gravity passes posterior and creates a counternutation movement [16]. The iliolumbar ligaments of the SI joint are mainly responsible for SI joint stability. The sacrotuberous, sacrospinous, and anterior sacroiliac ligaments also resist counternutation with lesser role in SI joint stability [1].

10.5.5 Muscles of the SI Joint

The muscles that correspond to the SI joint are commonly known as the pelvic floor muscles. These muscles provide stability to the vertebral column apart from their important function at the pelvis. The fibers of **levator ani** muscles form the floor of the pelvis. The muscle also has attachment to the urethra, sphincter, and walls of vagina. The **coccygeal muscle** is another important muscle in this region.

10.6 Pathomechanics of the Vertebral Column

The spine or the vertebral column is one of the most common segments of frequent injuries and biomechanical alternations leading to pain and discomfort. The presence of numerous structures and multiple attachments predisposes the segment to mechanical changes. We shall learn about the most common conditions here.

(a) **Spondylosis:** It is a degenerative condition of the spine due to repeated mechanical trauma and loading of the vertebral column. Very commonly seen at the cervical spine and known as the cervical spondylosis.

(b) **Spondylolisthesis:** The **pars interarticularis** of the vertebra that connects the body with posterior elements gets fractured and the superior vertebra slips over the inferior. It is very commonly seen at the lumbosacral junction **(L5–SI).** The reason can be attributed to the increase in the lumbosacral angle as seen with increased lumbar lordosis which in turn increases the anterior shear force. The vector diagram below explains the relationship of shear force with lumbosacral angle (Fig. 10.6). The normal lumbosacral angle has been reported as 30 ° [1].

 In the figure above, the line of gravity creates a resultant force vector represented as body weight (F). In line with the law of parallelogram, there would be two other forces perpendicular to each other. One force that acts to move the superior segment over the inferior due to the inclined angle is known as anterior shear component (Fs), whereas the other force which is perpendicular to the shear force is termed as the compressive force (Fc) and determines the magnitude of the compression on the vertebral bodies due to body weight. It can be

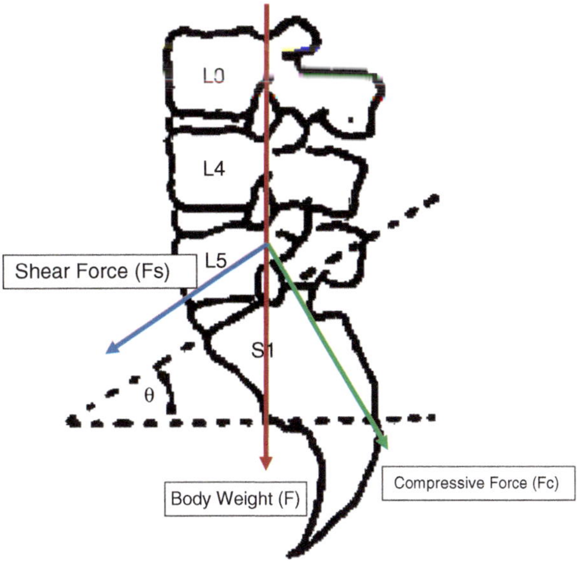

Fig. 10.6 Vector diagram for relationship of shear force and compressive force with lumbosacral angle. If lumbosacral angle increases the shear force would also increase

understood that an increase in the normal lumbosacral angle would cause greater shear force and higher chances of slippage [17].

(c) **Slipped disk/disk herniation/protrusion:** The slipped disk is related to movement of the nucleus pulposus outside the annulus fibrosus and end plate. This is commonly seen at the lumbar spine due to heavy weight lifting, bending movements, and altered mechanical loading. The most important measures for prevention of disk injury include (1) **strengthening of back extensor muscles** – the exercise should be performed in prone position so that muscles act against gravity and compressive force at the lumbar spine creates a negative pressure for the disk to suck in. Furthermore, strengthening would allow the spine to withstand higher forces without chances of failure, and (2) application of ergonomics—**stoop vs squat lifting (Fig. 10.7).**

In the figure above, we can see that stoop lifting involves bending of the thoracolumbar segments with the hip and knee straight, whereas the squat lifting is done with the knee and hip bent with straighter back. Biomechanically, stoop lifting is disadvantageous because the lever arm function of the back extensor muscles in this position is lost. The main muscle that can lift the weight from this position is back extensors, and we see that the muscles are completely stretched over the bent spine. Thus, the moments arm as well as the length tension relationship of the muscle is drastically affected (passive insufficiency) which will cause unwanted extra force application over the spine and thereby increase the chances of injury. In squat lifting, the leverage is transferred from the knee to the hip from where the back extensor can easily pull the load with optimal moment arm.

(d) **Altered curves of the spine:** The normal curves of the spine may get exaggerated as seen with **hyperlordosis** at the lumbar and **hyperkyphosis** at the tho-

Fig. 10.7 Depicts biomechanical comparison of stoop and squat lifting techniques

racic segment. In addition, the lordotic curve is poor as seen in the forward head neck posture in the cervical spine. The flexion and rotational changes in the vertebral column are known as **scoliosis** and very commonly seen due to poor posture.

(e) **Spinal stenosis:** Reduced size of the vertebral foramen leads to compression of the nerve in the canal and gives rise to a condition known as stenosis. This is commonly seen at the cervical and lumbar spine due to hyperextension.

(f) **Sciatica**: Altered biomechanics of lumbosacral spine can compress the sciatic nerve or nerve root which would lead to sciatica.

(g) **Pelvic floor dysfunction:** The pelvic floor muscles are very commonly seen to have weakened and lead to urinary incontinence.

10.7 Summary

In this chapter, we have learnt that the spine or the vertebral column is an important biomechanical segment which serves as the link between the upper half and lower half of the human body. Although segmental variations occur, the structure of the vertebral column allows efficient kinetic and kinematic functions of the soft tissues within and around. The vertebral column has six degrees of freedom and the kinematics of the cervical, thoracic, and lumbar segment has been summarized in Table 10.1.

Table 10.1 Summary of kinematics at the cervical, thoracic, and lumbar spine

Kinematics	Cervical spine	Thoracic spine	Lumbar spine
Osteokinematics—flexion	Sagittal plane and frontal axis	Sagittal plane and frontal axis	Sagittal plane and frontal axis
Arthrokinematics—flexion	Atlanto-occiput—occipital condyles roll forward and slide backward Lower cervical spine—anterior titling and gliding of superior vertebra	Anterior gliding and translation of superior vertebra over inferior (predominantly seen at lower segment)	Anterior gliding and translation of superior vertebra over inferior
Osteokinematics—extension	Sagittal plane and frontal axis	Sagittal plane and frontal axis	Sagittal plane and frontal axis
Arthrokinematics—extension	Atlanto-occiput—occipital condyles roll backward and slide forward Lower cervical spine—posterior titling and gliding of superior vertebra	Posterior gliding and translation of superior vertebra over inferior	Posterior gliding and translation of superior vertebra over inferior
Osteokinematics—lateral flexion	Frontal plane and sagittal axis	Frontal plane and sagittal axis	Frontal plane and sagittal axis
Arthrokinematics -lateral flexion	Coupled with ipsilateral rotation	Upper segment—coupled with ipsilateral rotation Lower segment—coupled with contralateral rotation	Coupled with ipsilateral rotation
Osteokinematics—side rotation	Transverse plane and vertical axis	Transverse plane and vertical axis	Transverse plane and vertical axis
Arthrokinematics—side rotation	Coupled with ipsilateral lateral flexion (atlanto-axial contributes 60%—pure axial rotation Lower cervical spine contributes 40%—coupled motion)	Distortion of ribs—convex posteriorly on ipsilateral side and convex anteriorly on contralateral side	Coupled with contralateral side flexion[a]
Range of flexion (in degrees)	0–60	0–50	0–60
Range of extension (in degrees)	0–75	0–45	0–25
Range of lateral flexion (in degrees)	0–45°	0–40	0–25
Range of side rotation (in degrees)	0–90°	0–30	0–18

[a]Coupling motion for axial rotation is different than cervical spine

References

1. Levangie PK, Norkin CC. Joint structure and function: a comprehensive analysis. Philadelphia, PA: F.A. Davis; 2011.
2. Kapandji IA. The physiology of the joints. In: The trunk and the vertebral column, vol. III. 2nd ed. Edinburgh London New York: Churchill Livingstone; 1974.
3. Cholewicki J, Crisco JJ III, Oxland TR, Yamamoto I, Panjabi MM. Effects of posture and structure on three-dimensional coupled rotations in the lumbar spine: a biomechanical analysis. Spine. 1996;21(21):2421–8.
4. Keller TS, Hansson TH, Abram AC, et al. Regional variations in the compressive properties of lumbar vertebral trabeculae: effects of disc degeneration. Spine. 1989;14:1012.
5. Crisco JJ III, Panjabi MM, Dvorak J. A model of the alar ligaments of the upper cervical spine in axial rotation. J Biomech. 1991;24(7):607–14.
6. Williams PL, Bannister LH, Berry MM, Collins P, Dyson M. In: Dussek JE, MWJ F, editors. Gray's anatomy. 38th ed. New York: Churchill Livingstone; 1995. p. 1382–4.
7. Nordin M, Frankel VH, editors. Basic biomechanics of the musculoskeletal system. Philadelphia: Lippincott Williams & Wilkins; 2001.
8. Moroney SP, Schultz AB, Miller JA, Andersson GB. Load-displacement properties of lower cervical spine motion segments. J Biomech. 1988;21(9):769–79.
9. Maiman DJ, Pintar FA. Anatomy and clinical biomechanics of the thoracic spine. Clin Neurosurg. 1992;38:296.
10. McGill S. Low Back disorders: evidence-based prevention and rehabilitation. Champaign, IL, Human Kinetics; 2002.
11. Neumann, DA. Kinesiology of the hip: a focus on muscular actions. J Orthop Sports Phys Ther. 2010;40(2):82–94.
12. Donald AN. Kinesiology of the musculoskeletal system. St. Louis, Mo: Mosby/Elsevier; 2002. isbn:978-0-323-03989-5.
13. Haher TR, O'Brien M, Dryer JW, Nucci R, Zipnick R, Leone DJ. The role of the lumbar facet joints in spinal stability. Identification of alternative paths of loading. Spine. 1994;19(23):2667–70.
14. Khoo BC, Goh JC, Lee JM, Bose K. A comparison of lumbosacral loads during static and dynamic activities. Australas Phys Eng Sci Med. 1994;17(2):55–63.
15. MacLaughlin SM, Oldale KNM. Vertebral body diameters and sex prediction. Ann Hum Biol. 1992;19(3):285–92.
16. Walker JM. The sacroiliac joint: a critical review. Phys Ther. 1992;72(12):903–16.
17. Kim HJ, Chung S, Kim S, Shin H, Lee J, Kim S, Song MY. Influences of trunk muscles on lumbar lordosis and sacral angle. Eur Spine J. 2006;15(4):409–14.

Kinetics and Kinematics of Hip and Pelvis

11

11.1 Introduction

The hip joint plays an important biomechanical link for the human lower limb. It is also known as the **Coxofemoral** joint. The joint allows mobility in all axis and planes in addition to providing stability for static and dynamic actions. The hip joint works in closed kinematic chain with the pelvis joint for an efficient biomechanical function. The muscles around the hip and pelvis are larger group of muscles with long moment arm and thus can produce power actions. The concentric and eccentric contraction of hip and pelvis muscles allow to withstand high compressive, tensile, and distraction forces. In comparison to the shoulder the hip joint complex is less mobile but more stable for being able to transfer the body weight efficiently. Let us understand the weight bearing functions of the hip joint in detail.

11.2 Anatomical Background

The hip joint is a synovial joint with ball and socket subtype. The joint is formed by the articulation between the head of the femur and the acetabulum fossa (Fig. 11.1).

11.2.1 Proximal Articulating Surface

Formed by the cavity/fossa of the acetabulum on the pelvis. This is a concave structure. The part of the acetabulum that articulates with the femur is covered by the hyaline cartilage and known as the lunate surface because of its shape. The **acetabular labrum** (wedge shaped fibrocartilage) deepens the articular surface and provides congruency to the joint. The congruency of any joint is determined as the maximum surface contact area. In other words the congruency to joint is its functional mobility with stability. The joint is most congruent in close pack position

A. Hazari et al., *Conceptual Biomechanics and Kinesiology*, https://doi.org/10.1007/978-981-16-4991-2_11

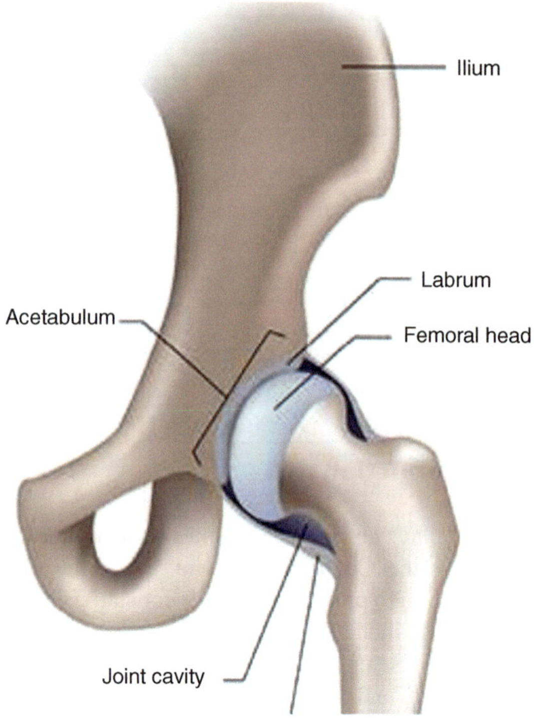

Fig. 11.1 Articulating structures of the hip joint

where the articular surface is in maximum contact whereas in the open pack positions the congruency is less. However at hip joint, the joint congruency is highest when it flexed, abducted and lateral rotation in a frog leg position. This position has been regarded as the true physiologic position of the hip joint [1, 2].

Functions of acetabular labrum:

1. Increase the contact surface area (Congruency) for the head of femur by deepening the fossa.
2. Proprioceptive role in protection of the acetabular rim. Experimental studies show that the labrum in hip does not have load bearing properties [2, 3].
3. Enhances joint lubrication by creating negative pressure under tight seal for the femoral head and mobilization of the synovial fluid.

The anatomical orientation of the acetabulum has important biomechanical significance. The acetabulum is placed **laterally, anterior, and inferiorly** which determines the congruency of the joint predominantly.

The extent of inferior tilt of the acetabulum forms the roof for covering the femoral head. The inferior tilt is referred as the **Center Edge angle** of the **Angle of Wiberg** [2]. On radiograph this is the angle formed between the line connecting the center of femoral head and lateral rim of acetabulum (Fig. 11.2a). The

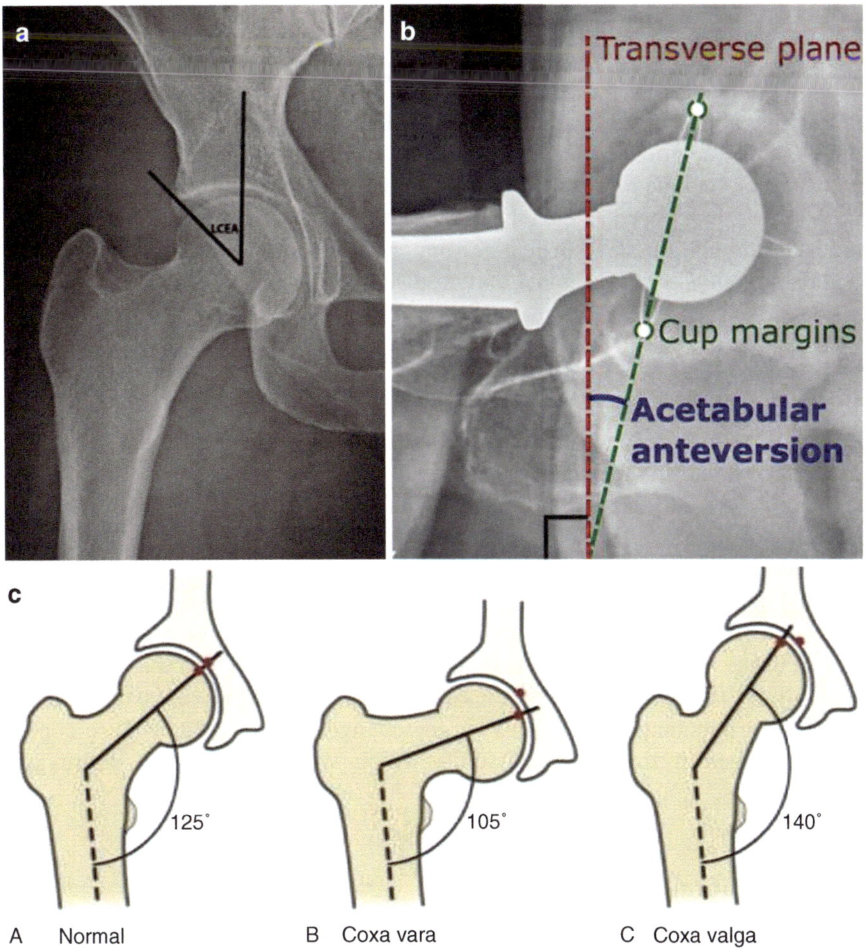

Fig. 11.2 (**a**) Showing central edge angle on radiograph. (**b**) Showing the acetabular anteversion angle on radiograph. (**c**) (**A**) Normal angle of inclination, (**B**) coxavara, (**C**) coxavalga

normal angle ranges from 22° to 42° (38 in men and 35 in women) as reported in the literature [2, 4]. A recently published article measured the lateral central edge angle (LCEA) among athletes and reported a normal range of 25°–45° [5].The children have lesser angle suggesting lower roof area of contact and thus the instability of the hip joint is more as evident by the higher chances of dislocation.

As we have learnt that the acetabulum is also placed anteriorly, the position gives anterior stability to the joint referred to as the angle of acetabular anteversion. The given angle is measured as the line connecting anterior and posterior acetabular margins and the line passing through the head of femur in transverse plane (Fig. 11.2b) [6].

11.2.2 Distal Articulating Surface

Formed by the head of the femur that is convex. It is covered by hyaline cartilage and is circular in shape. Like the orientation of acetabulum, the angulation of femur plays an important role in joint biomechanics. The head of the femur is placed superior, anterior, and medially to fit in the acetabular fossa. The proper orientation of head of the femur depends upon the alignment of the connection between the shaft and head through the neck of femur. Thus the head and neck of femur form two major angulation in respect to the shaft as discussed below.

11.2.3 Femoral Angle of Inclination

This angle is formed between the line passing through the neck of the femur and the shaft of femur in the frontal plane as we look anterior to posterior in anatomical position (Fig. 11.2a). The normal angle ranges from 115° to 140° [2, 7]. Since the pelvis of female is broader, a comparative lower angle of inclination is seen [2]. The alteration in the angle of inclination leads to pathological and biomechanical changes at hip. If the angle is increased, it is referred as the **coxa valga** and **coxa vara** if reduced than normal (Fig. 11.2b and c, respectively). The change in the normal angle would disturb the line of gravity and thus predispose the hip joint to higher degree of stress, wear, and tear due to excessive loading. In the close kinematic chain analysis, the change at hip would produce consequent changes at the knee and ankle which would disturb the entire gait biomechanics.

11.2.4 Femoral Angle of Torsion

The angle of femoral torsion is formed by the line passing through the femoral condyles and neck of femur in the transverse plane as we try to look from top to bottom in the anatomical position (Fig. 11.3). A normal angle of 10°–20° degree has been reported [2, 8]. An abnormal increase and decrease in the angle of torsion is known as the **anteversion** and **retroversion,** respectively. The condition that very commonly includes anteversion is Cerebral Palsy where the angle of 60° has been reported [9].

Steps to look for angle of torsion:

1. Hold the femoral bone in upside down below your head level and in front of the waist.
2. Look from the top at the femoral condyles through the neck of femur, we find that they both do not lie in the same plane. If we look carefully, the shaft of femur at midpoint is rotated slightly which produces the torsional angle.

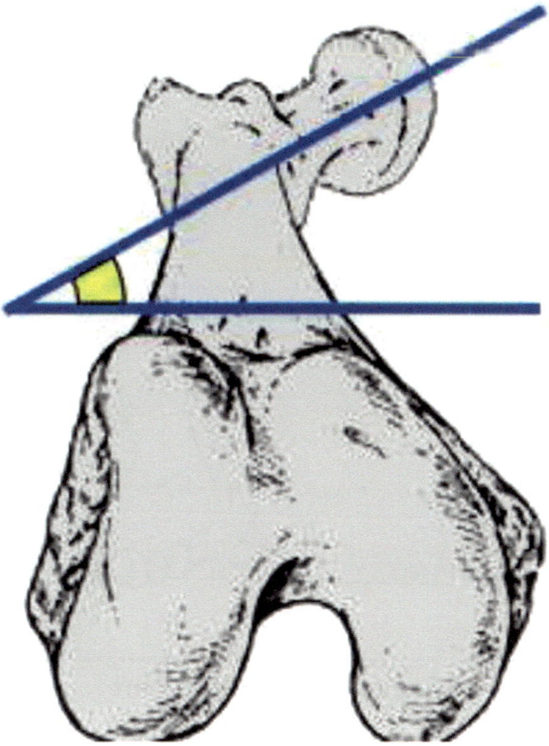

Fig. 11.3 Angle of femoral torsion

3. Point index finger of one hand through the head and neck of femur and the index figure of other hand through the femoral condyles to meet them. The angle formed would be the angle of femoral torsion.

Test for Femoral Anteversion The Craig's **test** for checking the altered angle of torsion has been used widely for clinical examination. This is also known as the **Trochanteric prominence angle test**. The patient is positioned in prone lying and with knee flexed to 90° on the test side. The greater trochanter is now palpated and the femur is rotated external direction to a position where the greater trochanter is most lateral and parallel to the bed. The angle between the starting position and final position produces the magnitude for angle of anteversion (Fig. 11.4).

Clinical Implication The change in the normal angulation of the hip would affect the biomechanics of the joint significantly. For instance, the angle would change the moment arm of the muscle and thus create lever arm dysfunctions. The muscles length tension relationship would be affected which would in turn affect the kinetic properties. In addition the kinematics like joint range of motion, line of progression, could be compromised. Squinting of patella and pigeon toe walking is a common clinical manifestation of increased femoral anteversion.

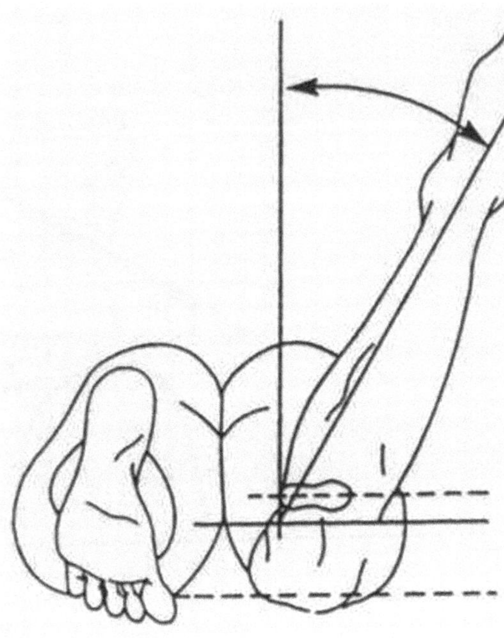

Fig. 11.4 Depicts clinical test for angle of anteversion (Craig's Test)

11.3 Kinematics of the Hip Joint

The hip joint has ability to move in all three axes and planes. The motions at hip include flexion, extension, adduction, abduction, medial and lateral rotation. Thus we see that the hip joint has six degrees of freedom and can translate as well as rotate. We shall learn them under the osteokinematics and arthrokinematic section as discussed below. We shall discuss the open chain kinematics first where the femur would move over the acetabulum. The close kinematic chain motion where the pelvis can move over the femoral head has been explained in the upcoming section.

11.3.1 Open Chain Kinematics

Flexion

Osteokinematics: occurs in sagittal plane and frontal axis.

Arthrokinematics: rotation is anterior and gliding is posterior as per the convex concave rule. However the flexion motion is seen as pure posterior spinning of femoral head [2].

Extension

Osteokinematics: occurs in sagittal plane and frontal axis.

Arthrokinematics: rotation is posterior and gliding is anterior as per the convex concave rule. However the extension motion is seen as pure anterior spinning of femoral head [2].

Abduction
Osteokinematics: occurs in frontal plane and sagittal axis.

Arthrokinematics: rolling upward/superior and gliding downwards/inferior.

Adduction
Osteokinematics: occurs in frontal plane and sagittal axis.

Arthrokinematics: rolling downwards/inferior and gliding upwards/superior.

Medial Rotation
Osteokinematics: occurs in transverse plane and vertical axis.

Arthrokinematics: rotation anterior and gliding posterior

Lateral Rotation
Osteokinematics: occurs in transverse plane and vertical axis.

Arthrokinematics: rotation posterior and gliding anterior

11.3.2 Close Chain Kinematics

The close chain kinematics of the hip is characterized by the movement of the pelvis (acetabulum) over the head of femur. If the femur is fixed as seen with weight bearing, the hip joint motion is facilitated by the movement of proximal segment (pelvis) and thus we consider this as the close chain kinematics. We shall learn about the available motions of the pelvis and the resultant motion of the hip joint in this section.

Anterior Pelvic Tilt
Osteokinematics: occurs in sagittal plane and frontal axis.

Arthrokinematics: anterior rotation and inferior gliding of the pelvic over femur (Concave on Convex). The resultant motion at hip joint would be flexion because the pelvis would move closer to femur resulting in decreased angle instead of femur moving close to the pelvis.

Posterior Pelvic Tilt
Osteokinematics: occurs in sagittal plane and frontal axis.

Arthrokinematics: posterior rotation and superior gliding of the pelvic over femur. The resultant motion at hip joint would be extension because the pelvis would move away from femur resulting in increased angle.

Lateral Pelvic Tilt: Unilateral Stance
Osteokinematics: occurs in frontal plane and sagittal axis.

Arthrokinematics: the lateral pelvic tilt is seen at unilateral stance when we take entire load on one limb and offload the other. This is usually seen in dynamic state (walking, running, climbing, etc.) but one can voluntarily do the activity in static state such as standing on one leg. For easy understanding let us assume that we are using stairs. We find that we put the entire load on one side of the lower limb at the lower step for stability and then try to lift the other limb to step higher. During the process, the pelvis goes for lateral pelvic tilt **(pelvic hike)** in respect to the load bearing limb. Similarly, while descending the stairs, one limb is stable over the higher step and the other limb offloads and descends down to the lower step which is referred as the **pelvic drop** in respect to the loaded limb. The pelvic hike and drop are translatory motion and during hike the offloaded limb glides superiorly and inferiorly while pelvis is dropped. In biomechanics, the pelvis hike and drop is also described as the adduction and abduction movements at hip as described below. Consider Fig. 11.5a, where the vertical lines represent the respective lower limb right and left, and the horizontal line represents the pelvis at 90° to both lower limb during bilateral standing. Now let the left lower limb be the loading limb over which the right limb translates superior for pelvic hike (Fig. 11.5b) and translates inferior

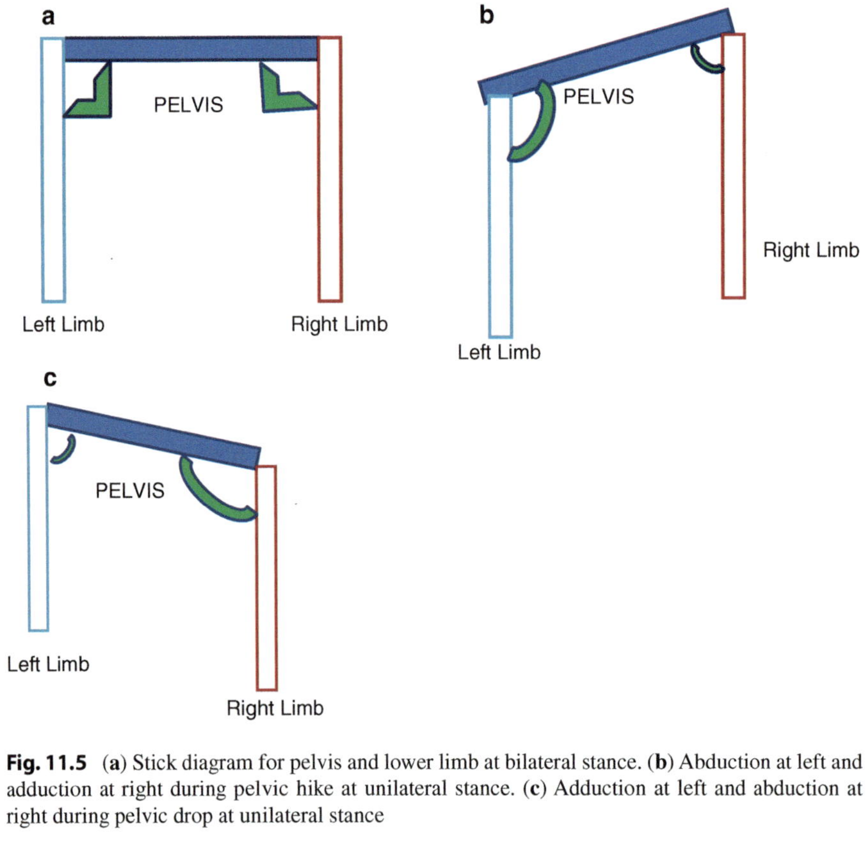

Fig. 11.5 (**a**) Stick diagram for pelvis and lower limb at bilateral stance. (**b**) Abduction at left and adduction at right during pelvic hike at unilateral stance. (**c**) Adduction at left and abduction at right during pelvic drop at unilateral stance

for pelvic drop (Fig. 11.5c) during unilateral stance. We find that during the pelvic hike, the angle between the left limb (represented by vertical line) and the pelvic (represented by the horizontal line) has increased more than 90° and thus it can be said that there is abduction of hip at the loading limb (ipsilateral). On the other hand, we find that the angle between the right limb and the pelvic represented by the vertical and horizontal lines, respectively, has reduced than 90°, thus there is adduction of hip at the contralateral limb.

Lateral Pelvic Tilt: Bilateral Stance

The lateral pelvic tilt in bilateral stance is seen when both limb are on the ground, however one limb is flexed at hip and knee (partial weight bearing) as seen with leaning on an object in standing position. Biomechanically, this can be seen as the pelvic drop on the side of the flexed lower limb. Thus the limb which is bearing weight partially would be abducted and other limb would be adducted as shown in Fig. 11.6. The lateral pelvic tilt in bilateral stance is an important biomechanical feature for normal gait patterns. The transfer of weight from one limb to other easily takes places with lateral shifts.

Pelvic Rotations: the pelvis has the ability to rotate forward and backwards which is termed as the anterior and posterior pelvic rotations, respectively.

Osteokinematics: occurs in transverse plane and vertical axis.

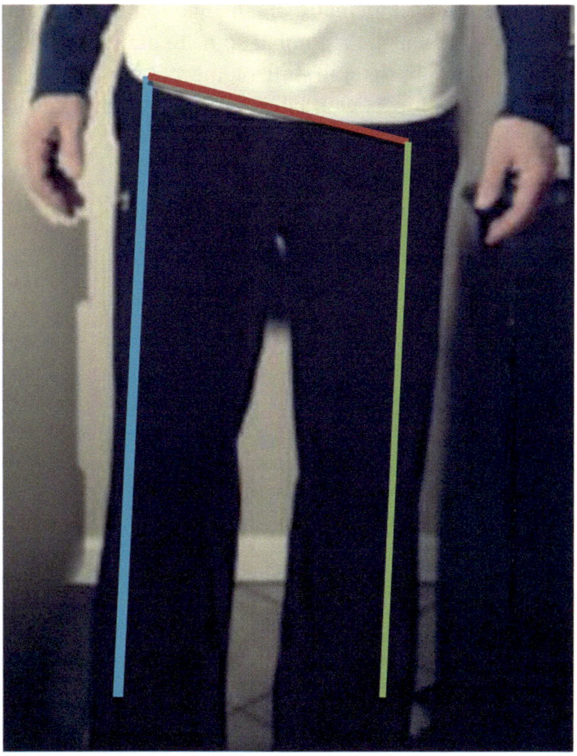

Fig. 11.6 Lateral pelvic tilt in bilateral stance. Note that the left hip is abducted as the angle is more than 90° and right side is adducted.

Arthrokinematics: involves the anterior rotation of one side of pelvis and posterior rotation of the other side. For instance, we are moving ahead during normal walking with left leg as leading limb, the right pelvis would rotate anteriorly and left pelvis would rotate posterior. Thus it is a reciprocal movement with shift in center of gravity from one to other limb [2]. In simple words, the opposite side of pelvis would rotate forward in respect to the leading lower limb during a gait cycle.

Open and Closed Kinematic Functions of the Pelvis, Hip, and Lumbar Spine
It is quite evident that the movement of lower segment would affect the upper segment in close kinematic chain. Thus the pelvis motion affects hip which in turn would affect the lumbar spine movements and continue till head. In this section, we shall restrict ourselves to the integrated action of pelvis, hip, and lumbar spine which is termed as the **pelvifemoral rhythm**. The pelvis and hip range of motion is exaggerated by the movement of lumbar spine in the pelvifemoral rhythm in an open chain movement as seen with touching the ground with straight arms and knee in bending forward position. The other example that depicts the pelvifemoral rhythm is abduction of hip in side lying position. The normal hip abduction range is 45°, however we can abduct above 80° in side lying position where the range of motion is adducted by the lateral pelvic shift and lateral flexion of the lumbar segment.

In the closed kinematic chain function the movement at pelvis joint creates a countermovement at the lumbar spine to maintain the center of gravity within limits. For example, the lateral shift of the pelvis would create adduction or abduction movements at hip as well lateral flexion on the lumbar spine as shown in Fig. 11.7 and Table 11.1.

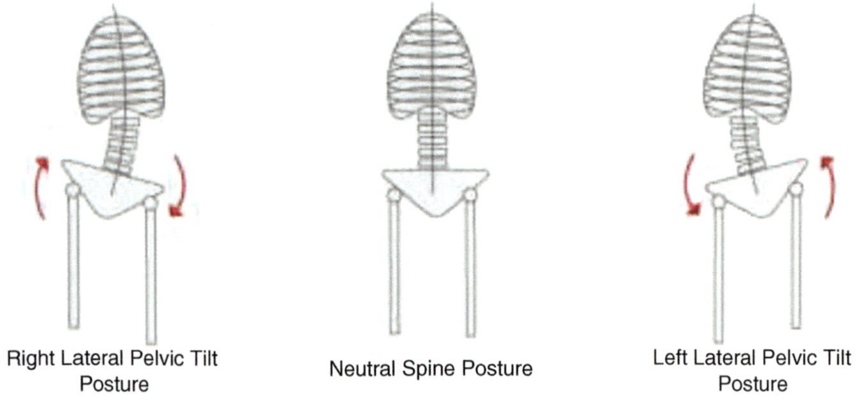

Fig. 11.7 Depicts closed kinematic function of pelvis. Hip and lumbar spine

Table 11.1 Integrated close kinematic function of pelvis, hip, and lumbar spine

Movement at pelvis	Associated movement at hip	Counter movement at lumbar spine
Anterior pelvic tilt	Flexion	Increased lordosis
Posterior pelvic tilt	Extension	Decreased lordosis
Lateral pelvic tilt (drop or hike)	Adduction or Abduction	Lateral flexion towards the adducted Hip irrespective of the hike or drop
Forward rotation	Medial rotation	Opposite side rotation
Backward rotation	Lateral Rotation	Opposite side rotation

11.4 Kinetics

The ligaments and joint capsules play an important role in stabilization and withstanding the forces of the hip and sacroiliac joints. Before we discuss the various type of forces or kinetics at these joint, it is important to know the ligaments that are present.

The most important ligaments of the hip joint include:

Ligamentum Teres: also known as the ligament of the head of femur and helps to stabilize the hip joint in semiflexion and adduction movements [2].

Iliofemoral ligament: it is the strongest ligament of the hip and also known as the Y ligament of Bigelow [2]. The ligaments help to stabilize the hip joint during hip extension motions.

Pubofemoral ligament: tensed under extension and prevents distraction of hip joint

Ischiofemoral ligament: weaker ligaments which get tensed under extension and prevent distraction of hip joint

11.4.1 Passive Stabilization Through Capsule and Ligaments

The hip joint ligaments and capsule are sufficient to stabilize the joint passively under the normal circumstance. In general the capsule and ligament complex can take two thirds of the compressive load due to body weight [2]. The line of gravity falls slightly posterior to the hip joint creating an external extension moment which is easily offset by the tension in the capsule and ligaments. If more force is required, the muscles come in picture and stabilize the joint dynamically.

We have also learnt that the congruency of the hip characterized by the maximum contact surface area or close pack position does not stand well for the hip. Instead the maximum congruency is seen as the frog leg position whereas as the close pack position would be extension, abduction, and medial rotation where the joint is comparatively congruent [2]. When the hip joint is neither in close pack position nor in frog leg positions the chances of hip instability are most as seen with flexion and adduction. Any small force from the lateral aspect on the shaft of femur can dislocate the joint easily and capsule and ligaments would fail to maintain the joint stability.

Clinical Significance of Capsuloligaments Complex It has been found that the tension in capsule and ligaments of the hip is least when it is flexed, abducted, and mid rotated. At this position, the intra-articular pressure is also minimized. Such an adaptive position of hip can be seen when a patient is suffering from pain due to problems in the capsule and ligaments or high intra-articular pressure.

11.4.2 Load Bearing Response of the Hip Joint

The hip joint has unique features to withstand the compressive and tensile loading imposed over it. The weight of the body from Head, Arm, and Trunk passes down to the acetabulum and head of femur through the pelvis. Therefore at the hip joint two distinct types of forces prevail as shown in Fig. 11.8.

We see that the weight of body and ground reaction force create a torque around the head of the femur. The torsional force due to body weight tends to

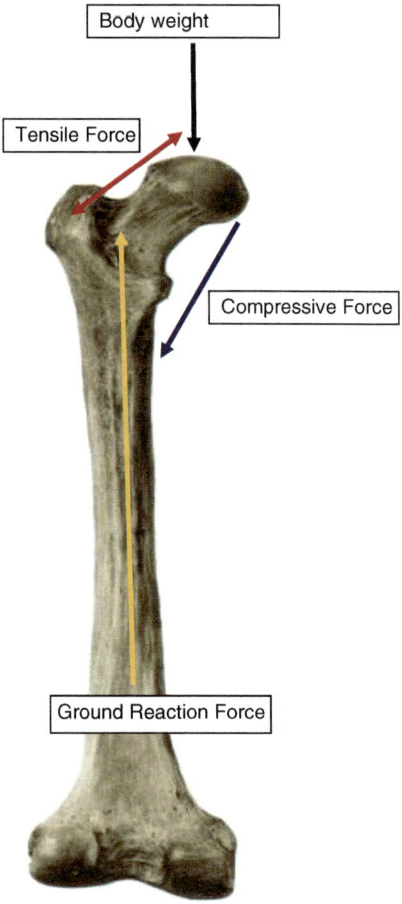

Fig. 11.8 Depicts compressive and tensile stress at the hip

push the head of the femur down and as a result the tensile force is high on the lateral side of the femoral shaft whereas the compressive force prevails on the medial aspect. The ability of the hip joint to withstand this stress is taken care by the cortical bone and their **trabeculae system** as suggested by Levangie and Norkin [2] (Fig. 11.9). The medial trabeculae are aligned vertically and therefore would resist the compressive loading. The lateral system are aligned horizontal, therefore it would resist the tensile force. If we look carefully, there is an area where the trabeculae are not present; this is the known as **zone of weakness** at which the **fracture neck of femur** is most common. It should be noted that the altered forces at the hip joint would lead to degenerative changes and therefore arthritic conditions at hip prevail.

11.4.3 Kinetics in Bilateral Stance

We have learnt that line of gravity lies posterior to hip joint and creates an external extension moment which is controlled by the passive tension in the capsule and ligaments. Thus the sagittal plane stability is passively maintained.

While we analyze the frontal plane stability and look at the force distribution in symmetrical bilateral stance, the weight of head, arm, and trunk would fall in between the legs and mid of the pelvis (acting as pivot). Thus equal weight shall be distributed to each hip joint which would create an equal and opposite gravitational torque as the force would be equal on both sides (weight of the body is equally

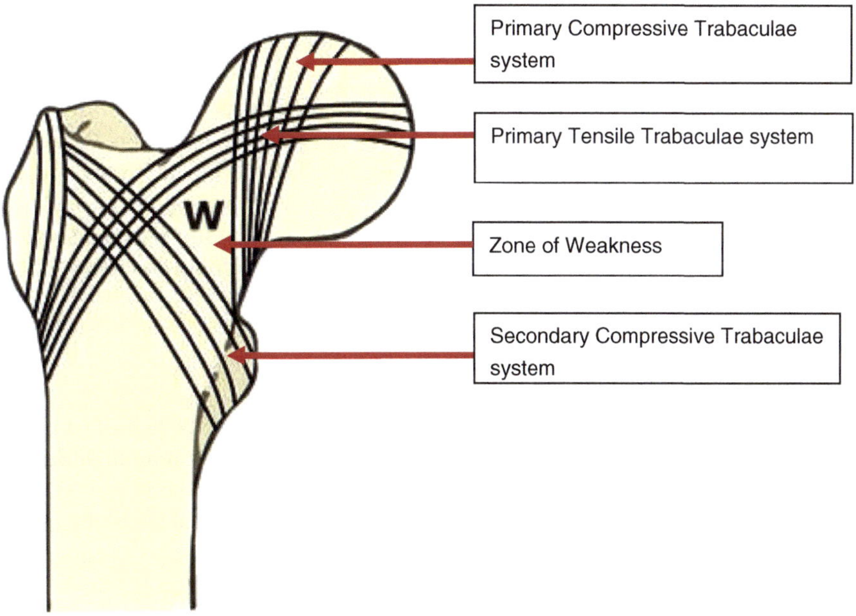

Fig. 11.9 Trabeculae system at hip joint

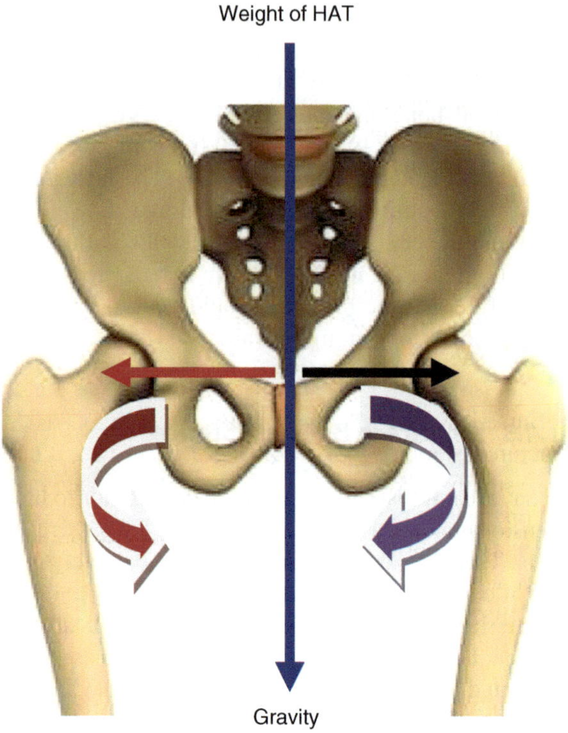

Fig. 11.10 Depicts the frontal stability at hip by equal and opposite gravitation torque

distributed and moment arm is same) as shown in Fig. 11.10. The force on the right and left cancel each other and stability is maintained without active contraction of muscles.

At instances where the bilateral stance is not symmetrical as seen with pelvic shift where one hip goes for adduction and the other goes for abduction, the active contraction of muscles would be required to stabilize the hip in frontal plane. Referring to Fig. 11.6, the abductors of the left hip needs to contract strongly to prevent excessive drop. In addition the adductors of the right hip would also contract to maintain the synergy and prevent overaction of the left hip abductors.

11.4.4 Kinetics in Unilateral Stance

During unilateral standing, the active contraction of hip abductors is of utmost importance to maintain the stability in frontal plane. The loading limb in unilateral stance has to maintain the superimposed weight of HAT (2/3 of body weight) and the weight of the offloaded limb (1/6 body weight) [2]. Let us calculate the hypothetical force that should be generated by the hip abductors in order to counter the gravitational torque in unilateral stance. The line of action for hip abductors

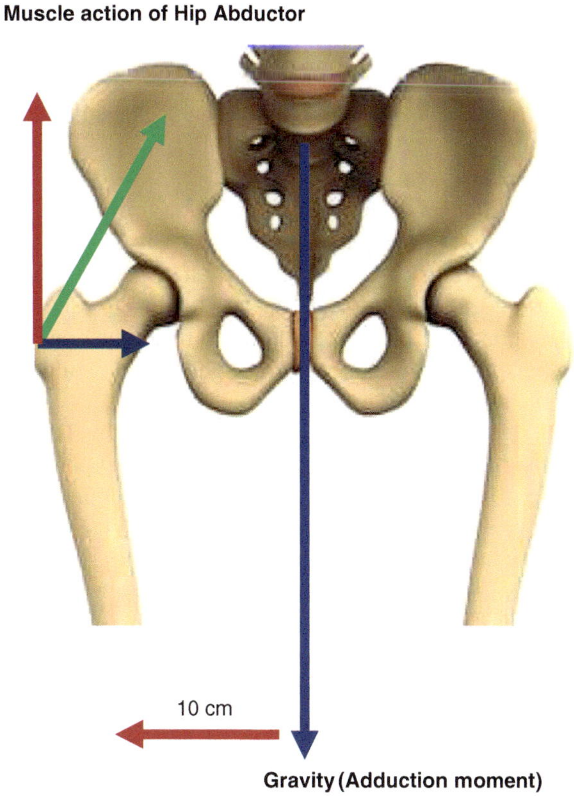

Fig. 11.11 Depicts abductor moment against the gravitation moment at hip joint

generates the force as represented in Fig. 11.11. The moment arm for the muscle is 5 cm and for the gravitation is 10 cm hypothetically [2].

Given,

Body weight = 300 Newton (N)

Weight of the unilateral hip = 2/3 × 300 + 1/6 × 300 = 250 N

Moment arm of abductor = 5 cm

Moment arm of Gravity = 10 cm

Force of Hip Abductor (FAB) = ?

Considering that the force required by the hip abductors should be equal and opposite to counter the external gravitation adduction torque, we can mathematically write the relation as equation below

$$\text{FAB} \times 0.05\,\text{m} = 200\,\text{N} \times 0.1\,\text{m}$$

$$\text{FAB} = 400\,\text{N}$$

Now the total compressive force at the loaded hip is equal to the body weight and the compressive force produced by the hip abductor. Thus in total during unilateral stance for a 300 N individual, the hip would experience a total compressive force = 400 + 250 = 550N.

It should be noted that, the calculation is hypothetical and factors like joint angle and force on the other limb have been neglected.

Clinical Implication

1. The concept of the above calculation forms the basis for **compensatory lateral leaning** at hip as seen during the pain. The reason that the individual leans is attributed to lesser force generation by the hip abductors against the gravitational torque. During the lean, the gravitational torque is reduced by bringing the line of gravity closer to the joint. As a result the gravitational adduction force would be less and therefore less force generation would be required to counter it by the hip abductors.
2. Use of Cane ipsilaterally Vs Contralaterally: The mechanical basis for prescribing the use of cane on the side of pain or opposite side totally depends upon the condition of the individual. If the pain is due to compressive loading and joint degeneration is suspected, the cane should be used contralaterally. The reason is that although the cane used on the side of the pain would reduce the body weight passing through the loading hip joint by 15% [2], the compressive loading is high in comparison to lateral leaning suggesting that a significant compression is still expected which could lead to further degeneration. On the other hand, if cane is used contralaterally, it would not support the body weight distribution but help to increase the moment arm of latissimus dorsi muscle from the joint center. The latissimus would hike the pelvis for a forward swung and thus hip abductors would produce less force to counter the gravitation torque which in turn would reduce the total compressive force.

11.5 Muscles of Hip Joint

The hip joint muscles provide dynamic stabilization to hip. These muscles are large and can generate a significant force. Muscle around the hip joint has a combination of one joint and two joint.

Flexors Primary muscles—**iliopsoas,** rectus femoris, Tensor fasciae latae (TFL)/ iliotibial band (IT)**,** and **Sartorius** [2]. Iliopsoas is the most important hip flexor which can also cause lumbar extension in close kinematic action. The role of IT band is important biomechanically as its dysfunction lead to snapping hip syndrome. Secondary hip flexors are the **pectineus, adductor longus, adductor magnus,** and the **gracilis** muscles [2].

Extensors Primary hip extensor includes gluteus maximum and hamstring (biceps femoris, semitendinosus, and semimebranosus) [2]. The gluteus maximum is the major muscle for hip extension and happens to be the largest. The moment arm is also very high with optimal length tension relationship throughout the range. The maximus also works eccentrically to control the overaction of quadriceps as seen during downhill walking. The hamstring is two joint muscles that have lower momentum arm in comparison to gluteus maximus. The hamstring experiences active insufficiency when the hip is extended and knee is flexed over 90°.

Adductors Muscles like **adductor brevis, adductor longus, adductor magnus, pectineus and gracilis** act as hip adductors [2]. The muscle has important role in controlling the force of abductors and stabilizing the hip.

Abductors **Gluteus medius** and the **gluteus minimus** are prime abductors [2]. Gluteus maximum is also a supporting abductor. We have learnt the role of hip abductors to withstand the compressive forces and frontal stability. The weakness of hip abductor often leads to lateral leaning which is also known as the **Trendelenburg gait.**

Lateral Rotators **Obturator internus** and **externus,** the **gemellus superior** and **inferior,** the **quadratus femoris,** and the **piriformis** are prime lateral rotators [2]. The line of action for quadratus femoris is perpendicular to the femoral shaft due to which they have good moment arm and angle of pull for strong rotation. Piriformis muscle lies parallel to the femoral head which make them good compressors and provide joint stability.

Medial Rotators The muscles of the hip that act as hip abductors and flexors also work as the medial rotators. The muscle of prime importance includes **gluteus medius, gluteus minimus, and the TFL.** It has been found that medial rotation torque increases with increased hip flexion [2, 10]. Thus the crouch gait could also be due to the result of hip flexors pathomechanics instead of adductor spasticity which contribute in hip medial rotation.

11.6 Pathomechanics

The hip joint is subjected to various pathologies related to altered biomechanics. The most commonly seen include:

Osteoarthritis One of the most common conditions at hip due to constant wear and tear of the joint capsule leading to moderate to severe degenerative changes. The altered forces at the hip are the major factor for developing the arthritic changes where the hip abductors play an important role.

Snapping Hip Syndrome The condition is also known as the IT band syndrome because the length tension relationship of the muscle is altered and IT band rubs over the greater trochanter in sagittal plane motion producing a snapping sound.

Piriformis Syndrome The condition is characterized by the compression of sciatic nerve passing under the **Piriformis** muscle due to tightness or altered hip joint biomechanics. The symptoms include radiating and tingling pain from the gluteus region down to toe at the dorsal aspect.

Fracture Neck of Femur The neck of femur is a very common site for fracture due to high torsional stress and deficiency of trabeculae system at the site of fracture.

Hamstrings Strain The hamstring strain is a common pathomechanics at hip where the muscle contracts or stretches violently leading to damage of the muscle fibers and sometimes tear. It is mainly seen among athletes or recreational players where the excessive stress over the muscles while contracting concentrically or eccentrically may lead to muscle damage or rupture.

Altered Gait due to Hip Pain The pain at hip joint could lead to altered gait biomechanics. The most commonly seen is the Antalgic gait, Trendelenburg gait, and Gluteus maximus gait. In the antalgic gait the stance time at the affected limb is reduced and the nonaffected side is loaded more. The gluteus maximus gait is characterized by the posterior leaning of the trunk at heel strike and thus known as the lurching gait. The trendelenburg gait is characterized by the weakness of the hip abductors leading to lateral leaning.

Lower Cross-Syndrome A very common musculoskeletal condition due to altered biomechanics at the pelvis, hip, and lumbar spine. The condition is characterized by the weakness of gluteus maximus and abdominals along with tightness of iliopsoas and back extensors. The prevalence rate is significant and higher in young female adults compared to men [11].

Sacroiliac Joint Dysfunction The SI joint is a common site of pain due to altered biomechanics. The more commonly known as the upslip and downslip of ilium at the SI joint causes pain, discomfort, and mechanical dysfunction.

11.7 Summary

The hip joint segment plays an important role in both open and close chain biomechanics. The joint is designed to take the load bearing function mainly. The ligaments and capsule of the joint are very strong and help to stabilize passively. The muscles around the hip joint are larger group of muscles with longer moment arm and thus produce high forces. The hip joint function is affected by the sacroiliac joint below and lumbar segment above. The open joint kinematics and kinetics of the hip joint has been summarized in Table 11.2.

Table 11.2 Kinematics and kinetics of hip joint

Hip joint motions	Osteokinematic	Arthrokinematics	Range of motion (Kendall et al.	Prime muscles
Flexion	Sagittal plane and frontal axis	Rotation is anterior and gliding is posterior	0–125°	Iliopsoas, rectus femoris, Tensor fasciae latae (TFL)/ iliotibial band (IT), and sartorius
Extension	Sagittal plane and frontal axis	Rotation is posterior and gliding is anterior	0–10°	Maximum and hamstring
Adduction	Frontal plane and sagittal axis	Rolling downwards and gliding upwards	From 45° abduction to 0°	Adductor Brevis, adductor longus, adductor magnus, pectineus, and gracilis
Abduction	Frontal plane and sagittal axis	Rolling upwards and gliding downwards	0–45°	Gluteus medius and the gluteus minimus
Medial Rotation	Transverse plane and vertical axis	Rotation anterior and gliding posterior	0–45°	Gluteus medius, gluteus minimus, and the TFL
Lateral Rotation	Transverse plane and vertical axis	Rotation posterior and gliding anterior	0–45°	Obturator internus and externus, the gemellus superior and inferior, the quadratus femoris, and the piriformis

References

1. Genda E, Iwasaki N, Li G, MacWilliams BA, Barrance PJ, Chao EY. Normal hip joint contact pressure distribution in single-leg standing—effect of gender and anatomic parameters. J Biomech. 2001;34(7):895–905.
2. Levangie PK, Norkin CC. Joint structure and function: a comprehensive analysis. Philadelphia, PA: F.A. Davis; 2011.
3. Petersen W, Petersen F, Tillmann B. Structure and vascularization of the acetabular labrum with regard to the pathogenesis and healing of labral lesions. Arch Orthop Trauma Surg. 2003;123(6):283–8.
4. Anda S, Svenningsen S, Dale LG, Benum P. The acetabular sector angle of the adult hip determined by computed tomography. Acta Radiol Diagn. 1986;27(4):443–7.
5. Tranovich MJ, Salzler MJ, Enseki KR, Wright VJ. A review of femoroacetabular impingement and hip arthroscopy in the athlete. Physician Sports Med. 2014;42(1):75–87.
6. Hingsammer AM, Bixby S, Zurakowski D, Yen YM, Kim YJ. How do acetabular version and femoral head coverage change with skeletal maturity? Clin Orthop Relat Res. 2015;473(4):1224–33.
7. Noble PC, Kamaric E, Sugano N, Matsubara M, Harada Y, Ohzono K, Paravic V. Three-dimensional shape of the dysplastic femur: implications for THR. Clin Orthop Relat Res. 2003;417:27–40.

8. Arnold AS, Delp SL. Rotational moment arms of the medial hamstrings and adductors vary with femoral geometry and limb position: implications for the treatment of internally rotated gait. J Biomech. 2001;34(4):437–47.
9. Heller MO, Bergmann G, Deuretzbacher G, Claes L, Haas NP, Duda GN. Influence of femoral anteversion on proximal femoral loading: measurement and simulation in four patients. Clin Biomech. 2001;16(8):644–9.
10. Delp SL, Hess WE, Hungerford DS, Jones LC. Variation of rotation moment arms with hip flexion. J Biomech. 1999;32(5):493–501.
11. Das S, Sarkar B, Sharma R, Mondal M, Kumar P, Sahay P. Prevalence of lower crossed syndrome in young adults: a cross sectional study. Int J Adv Res. 2017;5(6):2217–28.

Kinematics and Kinetics of Knee Joint 12

12.1 Introduction

The knee joint is one of the most complex joints of the human body. The joint serves the purpose of mobility and stability both. The structure and function of the joint make it suitable to perform dynamic movements with higher stability. Though the joint is biomechanically well stable, it is highly exposed to injury. There are multiple soft tissues in and around the joint which can lead to numerous joint dysfunctions. The weight of the body transferred from the hip is directly taken by the knee joint which impose high stress on the joint. However under the normal circumstances the joint is very strong and efficient.

12.1.1 Anatomical Background

The knee joint is a double condyloid joint consisting of the medial and lateral compartments. The joint is enclosed within a single capsule with two distinct articulations.

12.1.1.1 Articulations

(A) **Tibiofemoral articulation:** the knee joint is also known as the **tibiofemoral joint** where the **femoral condyles (convex)** are part of the proximal articulating surface and the **tibial plateau (flat surface)** forms the distal articulating surface (Fig. 12.1). The presence of menisci in between the bones enhances the joint congruency in addition to the structural compactness provided by intercondylar tubercles of tibia and intercondylar notch of the femur. Also the medial tibial plateau is larger than the lateral which enhances the articular surface during sagittal plane motion. The tibial plateau is anatomically inclined 7–10° posteriorly which is required for full range of flexion motion without early bony contact.

A. Hazari et al., *Conceptual Biomechanics and Kinesiology*,
https://doi.org/10.1007/978-981-16-4991-2_12

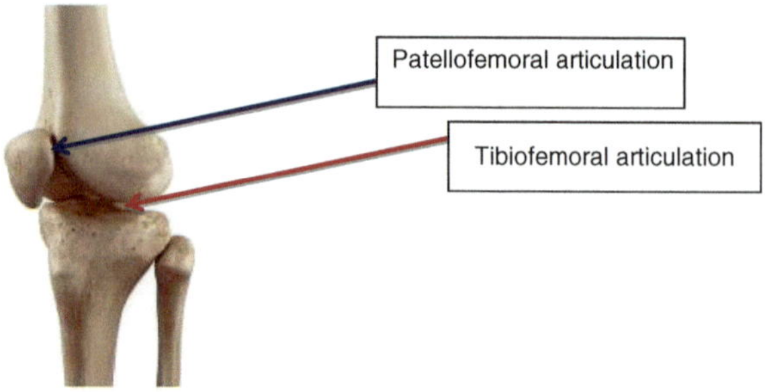

Fig. 12.1 Depicts articulations of the knee joint (tibiofemoral and patellofemoral)

(B) **Patellofemoral articulation:** The articulation of patella with femur is not a true joint rather it is a functional articulation without bone to bone contact. The patella is supported by soft tissues around and sits over the patellofemoral groove (Fig. 12.1).

12.1.1.2 Knee Joint Angulations

The mechanical axis or the weight bearing axis of the lower extremity passes through a line from the head of the femur and head of the talus [1, 2]. Thus we find that the axis is a straight line passing through the intercondyles and the normal angle for the tibiofemoral joint in the frontal plane is 180–185° [1] suggesting that the knee joint has physiologic valgus angle (5°). This is taken as the reference to describe the angulations at the knee joint. If the inner angle is increased beyond 185°, it is known as the **Genu Valgum** or **Knock Knees** (Fig. 12.2). If the inner angle is decreased, it is known as the **Genu Varum** or **Bow Legs** (Fig. 12.2).

12.1.1.3 Supporting Structures of the Knee Joint

The knee joint is supported by various structures like meniscus, capsule, ligaments, and muscles. In this section, we shall focus on the passive supporting structures as discussed below.

1. **Meniscus:** the menisci of the knee joint are reciprocal to disks at the vertebral column. These are firbrocartilaginous structure situated at the superior surface of the tibial condyles. The medial and lateral meniscus of the knee joint constitutes the major supporting structure for joint congruency. They form cavities where the femoral condyle sits and increases the contact surface area. The ability of the menisci to deform enhances the mobility and load bearing function of the joint very efficiently. The menisci have shown to withstand 50–70% of imposed body weight during dynamic motions [1, 3]. This is the reason they are also considered as the **shock absorber** of the knee joint. Let us understand the biomechanical function of menisci at the knee joint.
   (a) **Stability function:** the menisci of the knee joint enhance the contact surface area of the joint thereby increasing the joint congruency. If menisci were not present, the femoral condyle would sit on the flat tibial plateau which would cause high frictional force during the motion. Apart from that only a portion

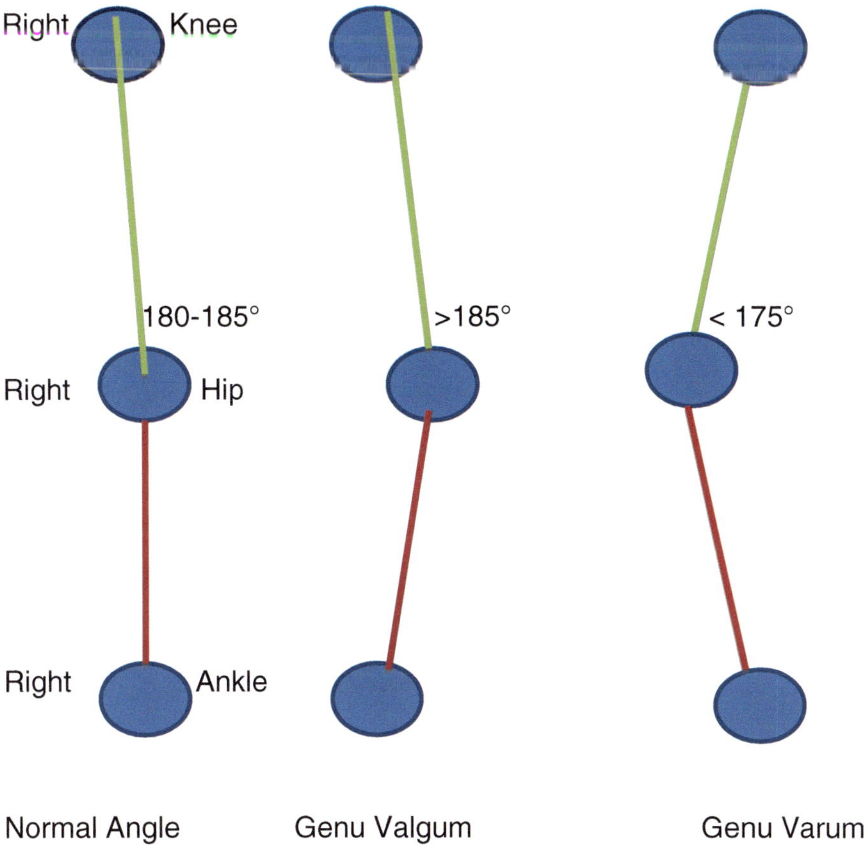

Fig. 12.2 Depicts genu varum and genu valgum at knee joint (refer the inner angle of the right side)

of the femoral condyle would form the articulating structure with the tibia because of its convexity. The presence of meniscus in between the surface thus enhances the contact surface area as shown in Fig. 12.3. Now, the movement of femoral condyle would produce less pressure over the tibia as the surface area is more (Pressure= force/ area). This would prevent wear and tear of the joint not allowing them for early degenerative changes. With aging, the meniscal function deteriorates and arthritis is common in elderly. Therefore this function of menisci at knee forms the basis of meniscal sparing during surgeries though these structures are painless. Only the periphery of the menisci contains nociceptors thus could lead to pain on tear.

(b) **Mobility function:** The knee joint kinematics or mobility is assisted by the meniscal shape and its ability to deform. If we closely observe the articulating surface of the knee joint, we find that the femoral condyle is bigger than the tibial plateau suggesting that the femoral motion over the tibial has to be simultaneous glide with rotation and a constant shift in the axis otherwise

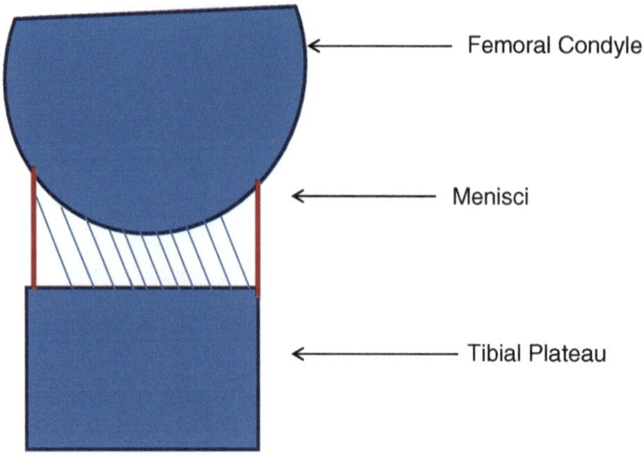

Fig. 12.3 Depicts meniscal function at the knee joint to enhance congruency

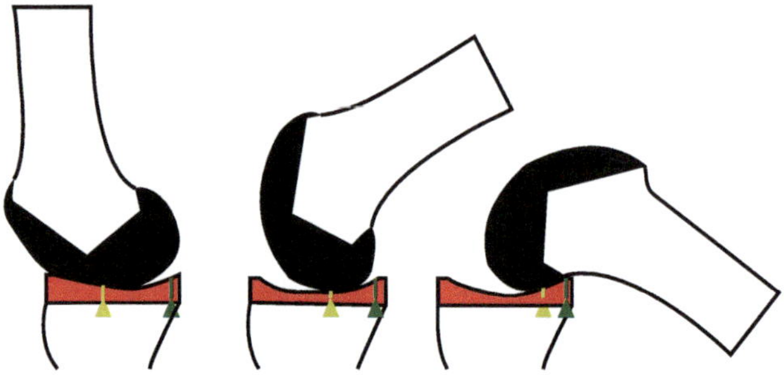

Fig. 12.4 Posterior deformation of the menisci and shift in axis of rotation during flexion at knee

the surface would be insufficient for the femoral condyles and they would slip off the tibial plateau. This principle of motion at the knee joint is facilitated by the wedge shape of the meniscus as shown in Fig. 12.4. In simple words, the wedge shape of the menisci allows the femoral condyles to ride over it during gliding and rotation [1]. Also the ability of the menisci to deform and remain intact between the articulating surfaces serves to move the joint very smoothly with less friction.

2. **Joint capsule:** The joint capsule of the knee joint is highly responsible to provide stability by enclosing the joint within the capsule. Thus the knee joint is intracapsular. The capsule restricts excessive motions through its attachments to ligaments and muscles. The capsule in the anteromedial and anterolateral aspects is well defined and known as the extensor retinaculum [1, 4]. In addition, the capsules have mechanoreceptors that help to stabilize the joint through reflex mediated muscular activation. Also the capsule forms a tight seal for the joint that allows efficient lubrication of the joint space by synovial fluid [1, 5]. The knee joint capsule consists of synovial and fibrous layers. The role of the synovial layers is to lubricate the joint and provide nutrition to avascular structures like menisci. The fibrous layers are superficial and form the retinacula of the knee joint to help stabilize it passively.

3. **Ligaments:** The ligaments of the knee joint are the most important passive stabilizers. These ligaments play a number of important roles and provide stability to joints in all planes depending upon their anatomical attachments. The most important ligaments of the knee joint include:
   (a) **Anterior cruciate ligament (ACL):** The ACL is the most popular ligament due to its frequent injuries and ability to cause severe dysfunction at knee. It provides stability to joint in the sagittal plane. The ligaments main function is to restrict the anterior translation of tibia over femur in close kinematic chain such as running, jumping, etc. [6]. The ligament has been shown to stabilize the varus and valgus stress secondarily [1, 7].

       The function of the ACL is supported by muscles on the posterior aspects of the lower limb such as Soleus and Hamstring. On the other hand, the strong contraction of Quadriceps would cause stress to ACL by pulling the tibia anteriorly.
   (b) **Posterior cruciate ligament (PCL):** The PCL restricts the posterior translation of tibia on femur and helps stabilize the joint in sagittal plane [1, 8]. In addition, it also provides stability to varus and valgus stress. Muscles like Popliteus and Quadriceps assist the action of PCL whereas the Hamstring and Gastrocnemius act as antagonist to PCL.
   (c) **Lateral collateral ligaments (LCL):** The LCL provides stability to joint in the frontal plane. There are two LCL present at the knee. The medial collateral ligaments are present on the medial aspect of the knee joint and resist the valgus stress whereas the lateral collateral ligament resists varus stress being present on the lateral aspect of the knee joint.
4. **Fascia:** the most functional and biomechanically important fascial attachment of the knee includes the iliotibial band (IT band). It is formed by the fascia of Tensor Fasciae Latae (TFL) and gluteal muscles and reinforces the anterior and lateral aspects of the knee [1, 9]. The IT band has been shown to assist ACL for preventing posterior translation of femur over tibia at knee extension [1, 9]. In addition, during flexion it assists the ACL by preventing anterior translation of tibia over femur. Thus we find that IT band is active and tensed in position of the knee joint. The IT band also plays an important role in patellofemoral joint function being attached to patella laterally. If the IT band is tight, it would lead to lateral patella displacement and interface with normal joint biomechanics.
5. **Bursae:** The bursae of the knee contribute to the joint biomechanics by lubricating the joint with synovial fluid during joint motion. The most important bursae of the knee joint include suprapatellar, subpopliteal, and gastrocnemius bursa.

12.2 Kinematics of the Knee Joint

The knee joint (tibiofemoral) has three degrees of freedom with predominant motion in the sagittal plane. The other available motions though lesser range at knee joint include medial and lateral shift (adduction and abduction) and medial and lateral rotation [1]. It should be understood that the axis for knee joint is not fixed and keep moving as we have shown in Fig. 12.4. This is important as the articular surface is very incongruent due to size difference between the tibial plateau and femoral condyle. In absence of the shifting or instantaneous axis the rolling and gliding would not take place efficiently.

12.2.1 Flexion

Osteokinematics: sagittal plane and frontal axis.

Arthrokinematics: For open chain kinematic where the tibia (concave) is moving over the femur (convex) such as high sitting and doing flexion or during swing phases of gait cycle), the tibia rotates as well as glides posteriorly.

Clinical Implication: In cases where knee flexion is restricted due to joint pathomechanics, posterior gliding should be given to increase the range of motion.

In close chain kinematics such as squatting where the femur is moving over the tibia, the flexion movement would include posterior rotation and anterior glide of femur (Fig. 12.5). At initial flexion range from 0 to 25°, there is pure posterior rotation of femur which is accompanied by anterior gliding beyond the 25° range of flexion at knee [1].

12.2.2 Extension

Osteokinematics: sagittal plane and frontal axis.

Arthrokinematics: in open chain kinematic where the tibia (concave) is moving over the femur (convex). For example; in high sitting on the bed and doing extension, the tibia would rotate as well as glide anteriorly.

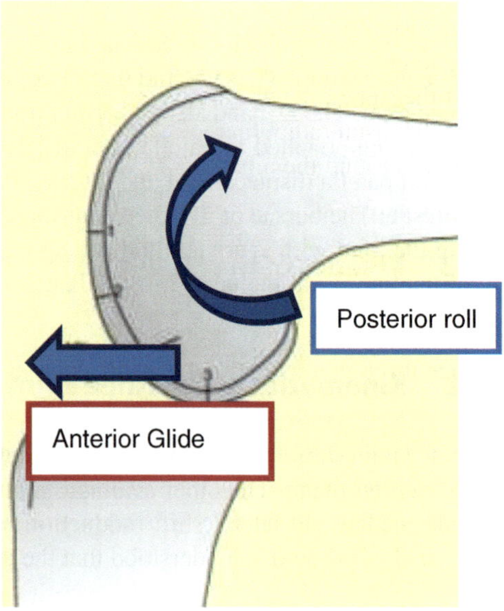

Fig. 12.5 Knee joint arthrokinematics during flexion in close chain

In close chain kinematics such as extension phase from squat to stand, the extension movement would include anterior rotation and posterior glide of femur. At initial extension range there is pure anterior rotation of femur which is accompanied by posterior gliding at mid and end ranges.

12.2.3 Medial Rotation

The medial rotation at knee joint is considered as the relative angular motion of the tibia with respect to femur. The medial tibial condyle is the pivot where the lateral tibial condyle moves through larger arc of motion compared to medial as explained below.

Osteokinematics: transverse plane and vertical axis.

Arthrokinematics: the medial tibial condyle rotates posteriorly slightly and the lateral tibial condyle rotates anteriorly through larger range of motion accompanied by the lateral rotation of the femur [1]. The maximum range of motion occurs at 90° knee flexion where the accessory structure is lax and there is maximum joint space for movements. The range of motion reduces as the knee joint approaches knee extension and found to be minimal at full extension.

12.2.4 Lateral Rotation

Osteokinematics: transverse plane and vertical axis.

Arthrokinematics: the medial tibial condyle rotates anteriorly slightly and the lateral tibial condyle rotates posteriorly through larger range of motion accompanied by the medial rotation of femur [1]. Similar to the medial rotation the range of lateral rotation is maximum at 90° knee flexion.

12.2.5 Adduction and Abduction of Knee

The knee abduction and adduction are frontal plane motion with limited range of motion. There is valgus and varus translation of tibia over femur resulting in abduction and adduction, respectively. The varus and valgus motions are essential coupled motion for flexion and extension of knee.

12.3 Locking Mechanism of Knee

The axis of motion for knee joint does not lie in the pure frontal plane for flexion and extension. The axis is oblique and presents closer to the frontal and deviated in sagittal plane slightly, thus flexion and extension motions are coupled with varus (adduction) and valgus (abduction), respectively.

In addition, the axis of rotation for knee joint is not fixed and keeps on moving as the **instantaneous axis** of rotation. During the terminal extension of knee in open chain kinematics, there is mandatory coupling of lateral rotation of tibia for bony locking of the knee joint which does not require any muscular force and thus it is known as the **automatic locking mechanism** of the knee. The reason for mandatory lateral rotation tibia is attributed to the difference in the size of medial and lateral tibial plateau and their kinematics as explained here.

Considering extension of tibia from full knee flexion, the motion starts at both tibial condyles. At 30° knee flexed position, the lateral tibial condyle, being smaller in size completes its rolling and gliding being smaller in size. Further extension is carried by the larger medial tibial condyle which leads to an obligatory lateral rotation of tibia predominately observed at the terminal extension (5°–0°) to attain the close pack position and expressed as the automatic locking kinematics of the knee [1].

In the event, the tibial tubercles are lodged in the intercondylar notch, the menisci are tightly interposed between the tibial and femoral condyles, and the ligaments are taut. Consequently, automatic rotation is also known as the **locking** or **screw home mechanism** of the knee [1]. The unlocking of the knee would start with medial rotation of tibia and initiated at the longer medial tibial plateau by Popliteal muscle. In load bearing close kinematic chain, the femur would rotate medially for terminal extension for locking unlike lateral rotation of tibia in open chain kinematics.

12.4 Kinetics of the Knee Joint

The knee joint is subjected to all type of loads including compressive by body weight, gravity and muscular contraction, shear and frictional force of joint, tensile force at ligaments, etc. The joint is biomechanically very efficient to withstand the forces and allows mobility with stability under the normal circumstances. The passive structures that support the joint have been dealt in Sect. 12.1.1.3 above. In this section, we shall learn about important muscles of the knee joint and their kinetic function.

12.4.1 Flexors

The muscles of the knee flexor group include **semimembranosus, semitendinosus, biceps femoris** (long and short heads), **sartorius, gracilis, popliteus, and gastrocnemius** [1].

Since flexion at knee is coupled with medial rotation of tibia, muscles like popliteus, gracilis, sartorius, semimembranosus, and semitendinosus have ability to cause flexion at knee. Also being attached medially these muscle can produce varum moments [1, 10]. The two joint hamstring muscles (semitendinosus, semimembranosus, and the long heads of the biceps femoris) are powerful knee flexors when the hip is flexed due to optimal length tension relationship. However when the hip is extended, the hamstring cannot produce strong knee flexion beyond 90° due to active insufficiency as it has become shortened at both joints. The contraction of hamstring creates posterior shearing force at tibia and thus makes it an agonist to ACL but antagonist to PCL.

Similarly, the gastrocnemius muscles will not be able to work as strong flexor of knee when the ankle is plantar flexed completely. This is the reason that gastrocnemius works as good knee flexor when ankle is neutral and knee is completely extended [1, 11]. Studies have shown that the flexor torque of the gastrocnemius reduces significantly with increased knee flexion [1, 11].

12.4.2 Extensors

The Quadriceps (rectus femoris, **vastus intermedius**, vastus lateralis, and vastus medialis) is the prime muscle for knee extension. Except the rectus femoris all are one joint muscle acting primarily at the knee joint. Studies reveal that the line of pull for **vastus intermedius** is parallel to the femoral shaft and thus considered as the **pure knee extensor**. The role of quadriceps is of prime importance to knee joint biomechanics. We shall discuss two scenarios where quadriceps action determines the basis for clinical implication.

(a) **Open vs. closed chain quadriceps's action:** In the nonweight bearing or open chain kinematics, the quadriceps would contract concentrically to create an extensor pull anterior and upwards (Fig. 12.6a). However during the weight bearing close kinematic chain, the eccentric contraction force of quadriceps can produce high compressive force at the knee joint as explained by the vector diagram in Fig. 12.6b.

    Clinical implication: Patients with knee joint arthritis should not be given squatting exercises as it would lead to high joint compression and damage the articular cartilage further.

(b) **Concentric vs. eccentric quadriceps training:** The quadriceps strength is very important to carry a variety of static and dynamic motion. The knee joint mobility and stability both depend upon the quadriceps strength. Therefore strength training for quadriceps should include the biomechanical concepts. The strength of any muscle can be increased by adding the external load in form of weights, gravity, etc., while choosing the appropriate plane and angle of pull.

Let us understand these concepts when we train the quadriceps muscle for concentric and eccentric contraction as explained in Fig. 12.7a and b, respectively.

When we apply an external load to the ankle joint and perform concentric contractions of quadriceps for knee extension, the same load creates different strength demand on the muscle against the gravity. We have learnt that the muscles need to generate an internal moment against the external gravitational moment to allow the motion in the desired plane of action. When the knee is flexed to 90°, the line of gravity passes close to the joint axis and thus quadriceps need to generate a small amount of force to initiate knee extension. With the same load applied, when the

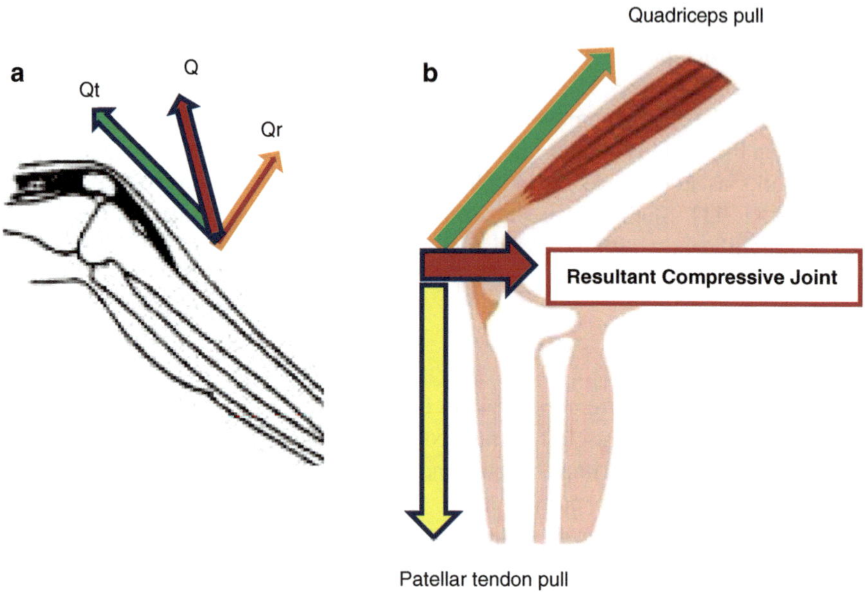

Fig. 12.6 (**a**) Depicts force vectors for quadriceps in nonweight bearing position creating and outward extensor pull (Resultant pull of quadriceps Q, translatory component Qt, and rotatory component Qr. (**b**) Depicts force vectors for quadriceps in weight bearing position creating compressive joint force

knee is extended gradually, the external gravitational moment increases as the distance (MA= moment arm) from the joint axis increases and therefore quadriceps need to generate more force to overcome the force of gravity (Fig. 12.7a). This is the reason that quadriceps is strongest at the initial ranges.

Similar concept is applied when we perform squats. However the quadriceps has to work eccentrically to stabilize the joint and overcome the external flexion moment created by the gravity. As the depth of quadriceps increases the eccentric contraction would need to be stronger as shown in Fig. 12.7b. Thus deep squats would produce more muscle strain and strength demand on the quadriceps compared to initial stages of squats.

12.4.3 Medial Rotators

The muscles of the knee that work as flexors also have the ability to rotate the tibia medially. The Sartorius and Popliteus muscle has been seen as the effective medial rotator at knee [1, 11]. The Popliteus is known as the muscle for unlocking although the knee joint medial rotation for flexion is an obligatory motion under the automatic mechanism and does not require muscular force.

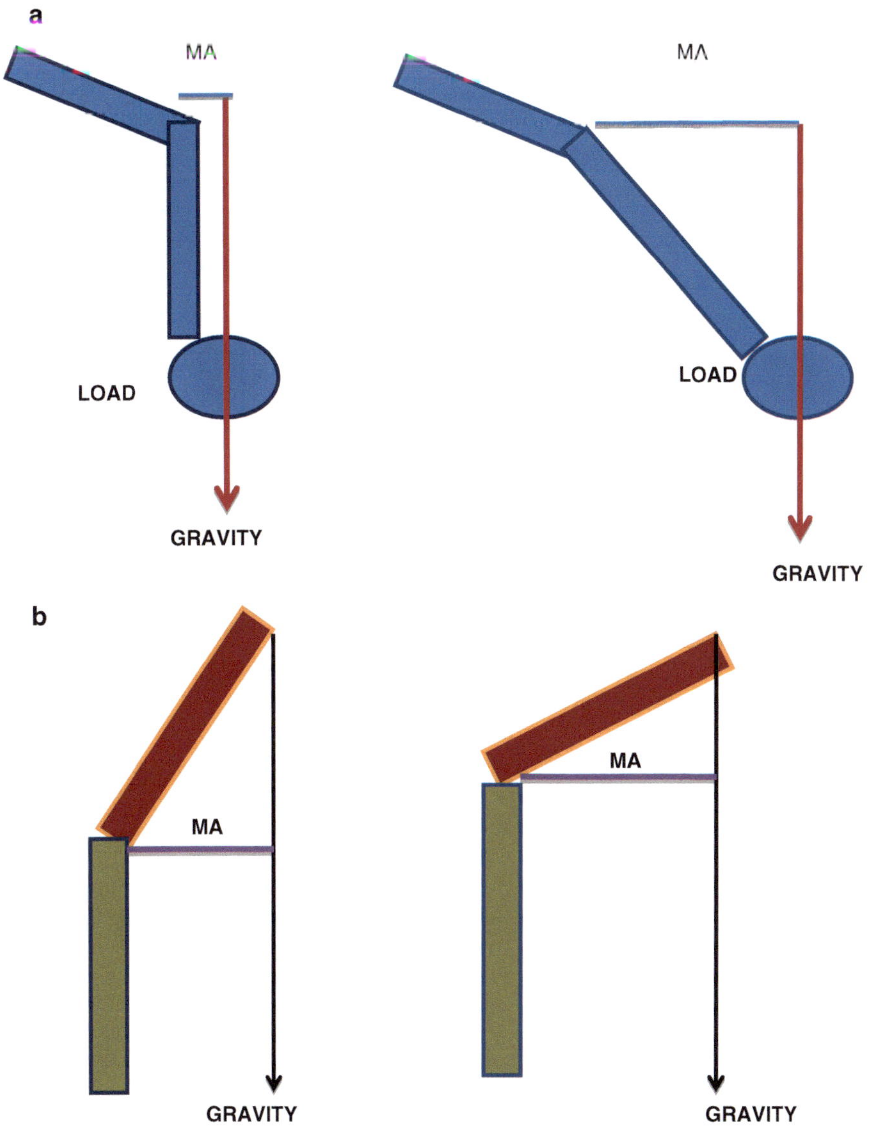

Fig. 12.7 (**a**) Depicts quadriceps concentric action against external loading and gravity. (**b**) Depicts quadriceps eccentric action during squatting against gravity. (**c**) Depicts patellofemoral joints motion

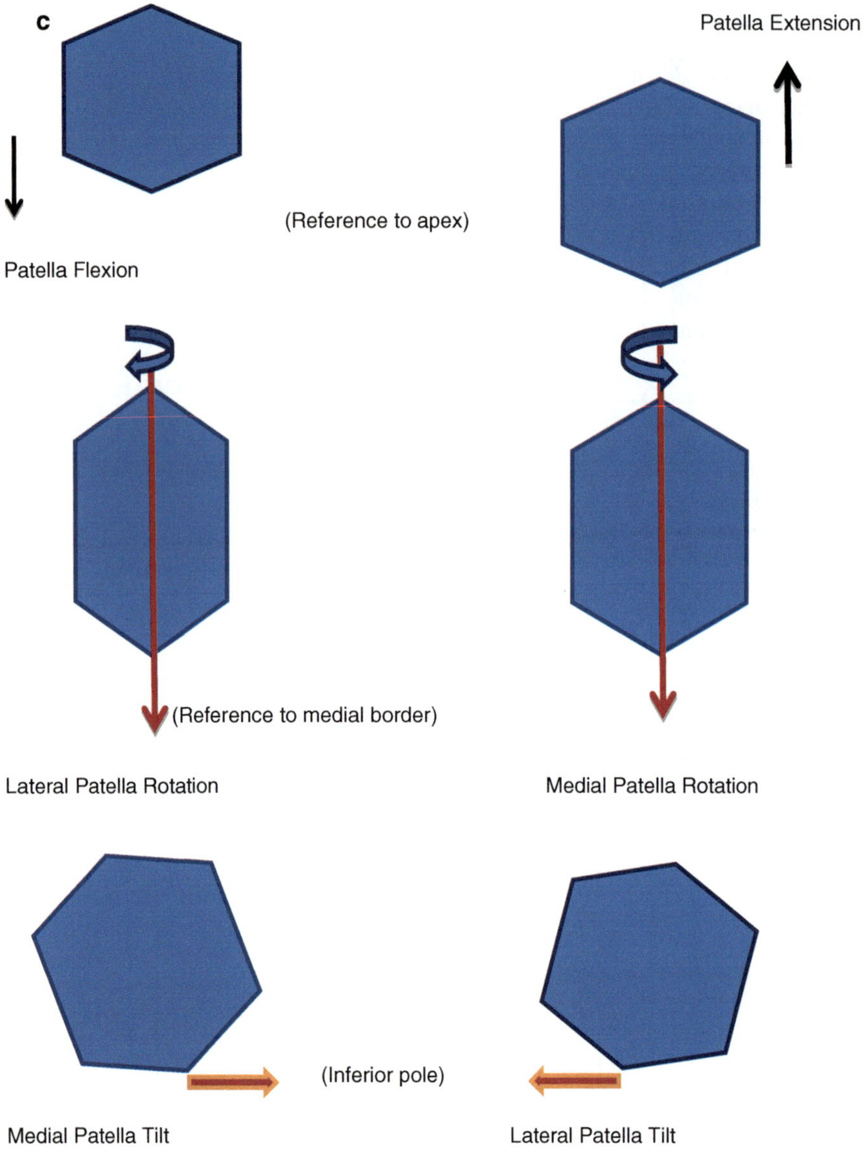

Fig. 12.7 (continued)

12.4.4 Lateral Rotators

The muscles of the knee that work as extensor also have the ability to rotate the tibia laterally.

12.5 Kinematics of the Patellofemoral Joint

The patellofemoral joints consist of articulation between the posterior surfaces of the patella over the anterior surface of the femoral condyles at knee extension. Inferiorly it is attached to the tibial tuberosity by the patellar tendon. The joint is functional rather a true synovial joint which mainly acts as the lever for knee joint kinematics. The kinematics of the patellofemoral joint is determined by the patellar ability to translate and rotate over the femur and tibia at different positions of knee joint. At knee extension, only the inferior pole of patella makes contact with femur [1, 12]. As the knee flexion progresses the articular contact of the patellofemoral joint also increases. The maximum articular contact is seen at 90° knee flexion [1, 13].

Let us now understand the motions available at the patellofemoral joint (Fig. 12.7c).

12.5.1 Patellar Flexion

During knee flexion, the patella tracks down over the femoral condyle and sits in the intercondylar groove at full flexion. The inferior tracking of patella is known as the patellar flexion. This is a sagittal plane motion and the movement of the patella should be considered in reference to its apex moving inferiorly.

12.5.2 Patellar Extension

The patellar extension is a relative motion from full knee flexion to knee extension where the patella translates upwards to reach its original position at full knee extension. This is a sagittal plane motion and the movement of the patella should be considered in reference to its apex moving superiorly.

12.5.3 Medial and Lateral Patellar Shifts

The patellar shifts take place in the transverse plane and vertical axis where the patella follows the femur in open chain kinematics. During knee flexion, the femur would rotate laterally which will cause lateral patellar shift and apposite could be seen during knee extension. The medial and lateral shift of the patella should be considered in reference to the border of the patella where the medial borders turn anterior and posterior around vertical axis and thus may be referred as patellar medial and lateral rotation also.

12.5.4 Medial and Lateral Patellar Tilts

The medial and lateral patellar tilts occur in the frontal plane where the patellar tends to follow its inferior attachment on the tibial tuberosity. During the knee flexion, the tibia rotate medially, therefore the inferior pole of the patella would tilt medially. The opposite would happen during knee extension. This motion should be considered in reference to the rotation of inferior pole of the patella.

12.6 Kinetics of the Patellofemoral Joint

12.6.1 Compressive and Joint Reaction Force

The patellofemoral joint is subjected to high joint compressive force compared to load bearing forces at the tibiofemoral joint. The patella acts as the anatomical pulley for the quadriceps muscle action which is responsible for modulating the patellofemoral joint forces. The ability of the patella to track in sagittal plane allows the quadriceps to increase in MA and maintain its length tension relationship throughout the knee range of motion. However the quadriceps force creates high compressive stress at the patellofemoral joint with increased knee flexion in weight bearing as explained in Sect. 12.4.2, point a and Fig. 12.6b. Consequently the joint reactions force would also be higher as per the Newton's third law. The other reason for increased joint compression force is small contact area of the patellefemoral joint. We have learnt that only the portion of patella makes contact with femur at different position of knee range of motion predisposing it to higher joint force. The altered anatomical position of patella can also have a significant influence on the patellofemoral joint kinetics as explained in the pathomechanics section ahead.

Clinical Implication: Isometric Quadriceps strengthening for arthritic patient is suggested with straight leg raise because at full knee extension the compressive force of the Quadriceps is minimal and contact surface area is also higher. At 90° knee flexion, the contact surface is more which helps to minimize the compressive force of the quadriceps and thus deep seated squats beyond 90° is also avoided.

12.6.2 Patellar Instability

We have learnt that the patellofemoral joint is very incongruent due to lesser articular contact. The instability of the patella is very common in frontal plane where the lateral patellar instability has been reported widely. The reason for frontal instability is that at full knee extension the patellar contact with the femoral sulcus is minimal. Apart, the compressive force of the quadriceps is also least. In addition, the physiologic valgus angle creates a resultant lateral pull on the patella which makes it very unstable and should be controlled by the longitudinal stabilizers which gives them medial and lateral stability in the frontal plane. The longitudinal stabilizers of the patella consists of the patella tendon, quadriceps compressive pull and patellotibial ligaments [1, 14]. In addition, the stability is also provided by transverse stabilizers such as the extensor retinaculum, patellofemoral ligament, and muscular stabilization by vastus medialis and lateralis, bony stabilization by larger lateral lip of femoral sulcus [1].

12.6.3 Clinical Assessment of Patellofemoral Joint Stability

The patellofemoral joint stability is very much dependent upon its compressive pull of the quadriceps and patellar tendon. Clinically, the stability of the patella can be measured by determining the angle of quadriceps pull which is known as the **Q angle**. It is the line connecting from Anterior Superior Iliac Spine (ASIS) to midpoint of patella and to the tibial tuberosity as shown in Fig. 12.8. The normal Q angle ranges from 10

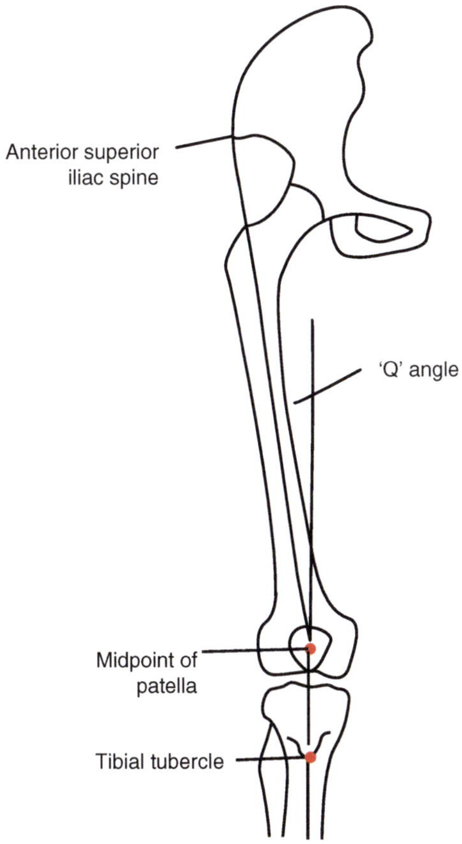

Fig. 12.8 Depicting Q angle and measurement of lateral pull of the patella

to 15° at full knee extension [1, 15]. An increase in the Q angle would increase the lateral force and instability to the patella. The common factors that increase the Q angle and subsequent lateral position of patella include muscular imbalance between vastus medialis and lateralis (lateralis being stronger will pull patella laterally), tight IT band, genu valgum, femoral anteversion, and lateral tibial torsion.

12.7 Pathomechanics

Due to the complexity of the knee joint, it is highly susceptible to altered biomechanics resulting from joint pathology, injuries, and aging process. A number of conditions can be listed. However we shall focus on the most important one from the biomechanical perspective.

(a) **Pathological angulations of knee:** Genu Varum and Genu Valgum are very commonly seen dysfunctions at the knee that could affect the joint kinetics and kinematic significantly. The Genu Varum would increase the compressive force on the medial structures including menisci, capsule, and ligaments whereas the lateral structure would experience increased tensile force. On the other hand, Genu Valgus would create compressive force on lateral structures and tensile force on the medial structure. The altered angle would also interfere with joint kinematics by affecting the available range of motion at the knee joint.

(b) **Fat pad and Plica syndrome:** fat pads of the knee are fatty tissues that support and cushion the knee joint. The infrapatellar fat pad also known as the **Hoffa's fat pad** is most common site of inflammation which leads to anterior knee pain. The Plicae are folds of synovium that fail to reabsorb during gestation. The Plicae when persist, contain loose, elastic tissues and get inflamed on irritation and affect the joint structure and function known as Plica syndrome [1].

(c) **Bursitis:** The chronic overuse injury to the knee joint can inflame the bursa of knee and cause biomechanical dysfunction. The main role of bursa is to mobilize the synovial fluid and provide nutrition to the joint.

(d) **IT band syndrome:** the IT band syndrome is characterized by the overuse injury to the tissue while consistently rubbing on the lateral aspect of knee during flexion and extension movements. The IT band syndrome can affect the lateral stability of the knee joint and cause pain, tenderness, etc.

(e) **Ligament injuries of the knee:** The ligaments of the knee are highly susceptible to injury because of their fast dynamic movements. In particular the ACL is commonly injured with semi flexion and twisting of lower limb. Studies report that at 30° knee flexion the injury chances are high because both the Anteromedial band (AMB) or Postero Lateral band of the ACL are loosened at this point [1, 16].

Apart from the ACL, the lateral collateral ligaments are also commonly injured and can compromise the collateral stability of the joint. A common injury involving the **medial meniscus, medial collateral ligament, and anterior cruciate ligament** is known as the **O'Donoghue triad.**

(f) **Patellofemoral pain syndrome:** the patellofemoral pain syndrome is characterized by the pain in the antero-lateral aspect of the knee which increases with activities involving sitting, climbing, squatting, etc. The pathoetiology of the condition is attributed to the high compressive force of quadriceps and joint reaction force which lead to degenerative changes. We have learnt how the compressive force at the knee joint is increased by strong quadriceps pull.

(g) **Knee osteoarthritis:** The degenerative changes at knee due to consistent wear and tear of cartilage are very commonly seen among elderly. The arthritis characterized by increased frictional force which leads to joint pain and bony spur formation.

(h) **Patella alta and baja:** the higher position of patella from normal is known as the Patella Alta, whereas the lower position is known as the Patella Baja. The ideal position of the patella is determined by the Insall–Salvati index which is the ratio of patella height to patella tendon and should be equal to 1 [17]. With patella Alta, the onset of contact between the quadriceps tendon and femoral condyle is delayed thus increasing the compressive force on the joint [1, 18].

(i) **Quadriceps lag:** We have learnt that the patella acts as a pulley in the lever system for the quadriceps tendon in sagittal plane motion. If the patella movement is affected or patella is removed surgically, the moment arm of the quadriceps would be affected and it will not be able to perform the knee extension completely in last 15°. This lag in the knee extension range is known as the Quadriceps lag.

(j) **Recurrent patellar dislocation:** the patellofemoral joint has higher degree of frontal plane instability. Any lateral force or altered biomechanics can displace the patella laterally and cause dislocation. The recurrent dislocation of patella is common among females and can occur even with normal walking.

12.8 Summary

The knee joint is biomechanically designed for both stability and mobility function. The complexity of the joint is required to maintain its structure and function. The passive structure like capsule and ligaments support the joint efficiently. Active structures like muscles play important role in dynamic stabilization. The important kinematic and kinetic features of the tibiofemoral joint in open chain are summarized in Table 12.1.

Table 12.1 Open chain kinematics and kinetics of knee (tibia over femur)

Available Motion	Osteokinematics	Arthrokinematics	Range of motion [1]	End feel	Muscles
Flexion	Sagittal plane Frontal axis	Posterior Roll and Posterior Glide of Tibia over femur	0–140°	Soft tissue approximation	Hamstring sartorius, gracilis, popliteus, and gastrocnemius
Extension	Sagittal plane Frontal axis	Anterior Roll and Anterior Glide of Tibia over femur	140–0°	Bony	Quadriceps
Medial rotation	Transverse plane Vertical axis	Medial tibial condyle rotates posteriorly, lateral tibial condyle rotates anteriorly	0–15°	Bony	Sartorius and Popliteus
Lateral rotation	Transverse plane Vertical axis	Medial tibial condyle rotates anteriorly, lateral tibial condyle rotates posteriorly	0–20°	Bony	Vastus lateralis
Adduction	Frontal plane Sagittal axis	Varus translation of tibia coupled with flexion	0–13°	Bony	Flexors
Abduction	Frontal plane Sagittal axis	Valgus translation of tibia coupled with extension	13–0°	Bony	Extensors

References

1. Levangie PK, Norkin CC. Joint structure and function: a comprehensive analysis. Philadelphia, PA: F.A. Davis; 2011.
2. Johnson F, Leitl S, Waugh W. The distribution of load across the knee. A comparison of static and dynamic measurements. J Bone Joint Surg. 1980;62(3):346–9.
3. Messner K, Gao J. The menisci of the knee joint. Anatomical and functional characteristics, and a rationale for clinical treatment. J Anat. 1998;193(2):161–78.
4. Starok M, Lenchik L, Trudell D, Resnick D. Normal patellar retinaculum: MR and sonographic imaging with cadaveric correlation. AJR Am J Roentgenol. 1997;168(6):1493–9.
5. Ralphs JR, Benjamin M. The joint capsule: structure, composition, ageing and disease. J Anat. 1994;184(Pt 3):503.
6. Petersen W, Tillmann B. Structure and vascularization of the cruciate ligaments of the human knee joint. Anat Embryol. 1999;200(3):325–34.
7. Shapiro MS, Markolf KL, Finerman GA, Mitchell PW. The effect of section of the medial collateral ligament on force generated in the anterior cruciate ligament. J Bone Joint Surg Am. 1991;73(2):248–56.
8. Ritchie JR, Bergfeld JA, Kambic H, Manning T. Isolated sectioning of the medial and posteromedial capsular ligaments in the posterior cruciate ligament-deficient knee. Am J Sports Med. 1998;26(3):389–94.

9. Recondo JA, Salvador E, Villanúa JA, Barrera MC, Gervás C, Alústiza JM. Lateral stabilizing structures of the knee: functional anatomy and injuries assessed with MR imaging. Radiographics. 2000;20(suppl_1):S91–S102.
10. Zhang LQ, Xu D, Wang G, Hendrix RW. Muscle strength in knee varus and valgus. Med Sci Sports Exerc. 2001;33(7):1194–9.
11. Li L, Landin D, Grodesky J, Myers J. The function of gastrocnemius as a knee flexor at selected knee and ankle angles. J Electromyogr Kinesiol. 2002;12(5):385–90.
12. Komistek RD, Dennis DA, Mabe JA, Walker SA. An in vivo determination of patellofemoral contact positions. Clin Biomech. 2000;15(1):29–36.
13. Nakagawa S, Kadoya Y, Kobayashi A, Tatsumi I, Nishida N, Yamano Y. Kinematics of the patella in deep flexion: analysis with magnetic resonance imaging. JBJS. 2003;85(7):1238–42.
14. Desio SM, Burks RT, Bachus KN. Soft tissue restraints to lateral patellar translation in the human knee. Am J Sports Med. 1998;26(1):59–65.
15. Livingston LA, Mandigo JL. Bilateral within-subject Q angle asymmetry in young adult females and males. Biomed Sci Instrum. 1997;33:112–7.
16. Torzilli PA, Greenberg RL, Insall J. An in vivo biomechanical evaluation of anterior-posterior motion of the knee. Roentgenographic measurement technique, stress machine, and stable population. JBJS. 1981;63(6):960–8.
17. Schlenzka D, Schwesinger G. The height of the patella: an anatomical study. Eur J Radiol. 1990;11(1):19–21.
18. Singerman R, Davy DT, Goldberg VM. Effects of patella alta and patella infera on patellofemoral contact forces. J Biomech. 1994;27(8):1059–65.

Kinematics and Kinetics of Ankle and Foot Complex

13

13.1 Introduction

The human ankle and foot complex consists of 26 bones, 33 joints and more than 100 muscles, tendons and ligaments [1]. The ankle complex is predominantly a weight-bearing joint and acts as a foundation for stability. Foot structures absorb the shock and transfer the body weight onto the ground thereby handling tons of force every day [2]. An adult walks 4000–6000 steps per day on an average [3]. We can expect biomechanical changes in the foot owing to the amount of weight and stress we transfer to our feet each day [4]. The ankle joint complex is composed of the foot and the lower segment of the leg, which acts as kinetic linkage and allows the lower limb to interact with that of the ground [5]. The ankle joint complex bears the high level of compressive and shears forces during walking [6]. The bony structures and ligaments allow the ankle complex to function with a higher degree of stability as compared to other joints of the lower limb. The ankle joint is less likely to undergo degenerative changes such like osteoarthritis unless a history of prior trauma is present [7]. The modern human foot anatomy has an interesting evolutionary story that determines the basis for drastic differences between our feet and those of our closest living animals, like apes. Several unique features of the human foot, including a spring-like longitudinal arch and short toes, are likely adaptations to long-distance running. Early human had an evolving feature of incipient longitudinal arch, adducted hallux, plantigrade foot posture; stiff midfoot during the push-off phase makes it convenient for bipedal walking [8].

13.1.1 Anatomy of the Ankle Joint

The ankle joint complex is composed of 26 bones and is functionally divided into three distinct segments such as forefoot, midfoot, and hindfoot [9–11]. The forefoot consists of 14 pharynx and 5 metatarsals; the midfoot is formed by 5 tarsal bones

A. Hazari et al., *Conceptual Biomechanics and Kinesiology*,
https://doi.org/10.1007/978-981-16-4991-2_13

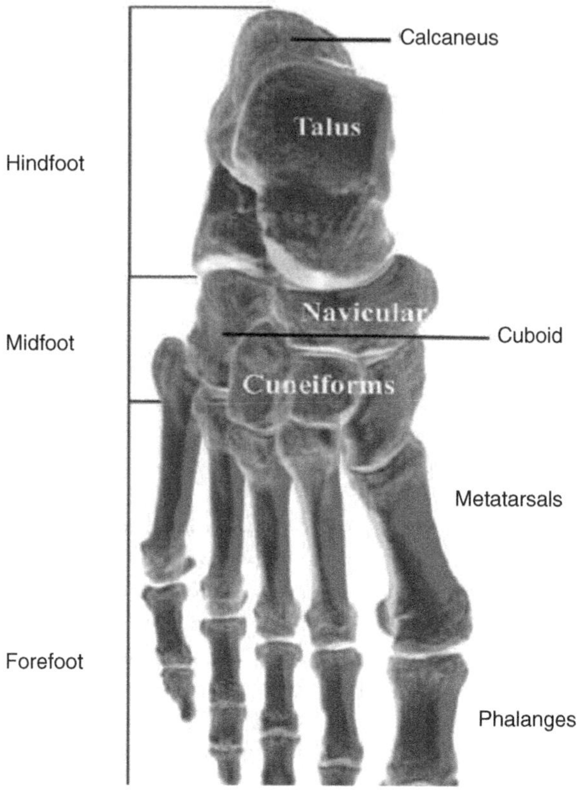

Fig. 13.1 Depicts segments and bones of foot and ankle complex

and the hindfoot is constituted by 2 larger bones like the talus and calcaneum as shown in Fig. 13.1.

13.1.2 Articulation of Ankle Joint

The talus articulates with the tibia; thus it is also known as the **talocrural joint.** In this book, the ankle joint would refer to the talocrural articulation of the hindfoot in particular. The talocrural joint is functionally linked to the **tibiofibular joint,** and thus, they are considered under the ankle joint biomechanics. It should be noted that the **proximal and distal tibiofibular joints** are part of the ankle complex and not the knee.

Proximal Articular Surfaces: Mortise formed by concave surface of the distal tibia and medial and lateral malleoli.

Distal Articular Surface: Formed by the body of the talus (convex).

13.1.2.1 Capsule and Ligaments

The capsule of the ankle joint is fairly thin and especially weak anteriorly and posteriorly [12]. The proximal and distal tibiofibular joint is supported by the superior and inferior tibiofibular ligament and interosseous membrane, respectively [12]. The important ligaments present at the talocrural/ankle joint constitute:

Crural tibiofibular interosseous ligament

Anterior and posterior tibiofibular ligaments

Medial collateral ligament (**MCL**)—known as the **deltoid ligament.** It is fan-shaped ligament and very strong. The main function is to check the valgus force at the ankle and restrict excessive calcaneal eversion.

Lateral collateral ligament (**LCL**)—The LCL helps control varus stresses and calcaneal inversion. It is composed of three separate bands.

Anterior talofibular ligament (**ATFL**)

Posterior talofibular ligament (**PTFL**)

Calcaneofibular (**CFL**)

The ligament is weaker than the MCL and often prone to ankle sprain during ankle twists in plantarflexion and medial rotation position. E.g., stepping upon a stone and twisted ankle.

13.1.2.2 Angulation-Biomechanical Axis of the Talocrural Joint

Normal Tibial torsion =30°[12].

In the frontal plane, the biomechanical axis for the ankle joint is inferiorly titled by 14°, whereas in the transverse plane, it lies 22° posteriorly and thus favors inversion over eversion range of motion and plantarflexion over the dorsiflexion.

13.2 Ankle Joint Kinematics

The major motion available at the ankle joint or hindfoot is **dorsiflexion and plantarflexion.**

13.2.1 Osteokinematics

The dorsiflexion and plantarflexion at the ankle joint take place in the sagittal plane through the frontal axis. The normal range of motion for dorsiflexion and plantarflexion is 10–20°, 20–50°, respectively.

13.2.2 Arthrokinematics

For nonweight-bearing joint motion, the convex talus moves within the mortise over the concave distal tibia and thus, rolling and gliding would be taking place in the opposite direction. For weight-bearing motions where the tibia would move over the talus, rolling, and gliding would take place in the same direction.

13.3 Kinetics at Ankle

The enhanced stability at the ankle joint in dorsiflexion allows the ankle to withstand compression forces of as much as 450% of body weight [12]. The **Gastrocnemius and Soleus** are prime plantar flexors of the ankle joint and are commonly known as the calf muscle. The Gastrocnemius is a two joint muscle. The tension in the anterior muscles like the **tibialis anterior, extensor hallucis longus, and extensor digitorum longus muscles** is the primary limit to plantarflexion. The posterior muscles like **Tibialis posterior, flexor hallucis longus and flexor digitorum longus** muscles help protect the medial aspect of the ankle, whereas the **peroneus longus and peroneus brevis** muscles protect the lateral aspect of the ankle joint.

13.4 The Subtalar Joint

The subtalar joint is the articulation between the talus and calcaneum and is therefore also known as the **talocalcaneal joint.** The major function of the subtalar joint includes dampening the rotational forces imposed by the body weight while maintaining stability at the ground. The subtalar joint consists a funnel-shaped tunnel known as the **Tarsal Tunnel**. The larger end is present on the lateral aspect towards the fibular malleoli and known as the **sinus tarsi,** whereas the smaller end is called **sustentaculumtali** present on the medial aspect. The tarsal tunnel could be a source of pain and ankle joint when structures passing through the tunnel get compressed.

13.4.1 Ligaments of the Subtalar Joint

A number of ligaments have been suggested at the Subtalar joint, as listed below [12].

- The calcaneofibular ligament.
- Lateral talocalcaneal ligament.
- The cervical ligament (Strongest-lies in the anterior sinus tarsi and joins the *neck of the talus to the neck of the calcaneus*).
- Interosseous talocalcaneal ligament.

13.4.2 Subtalar Joint Angulations

The oblique biomechanical axis for the subtalar joint makes it a triplanar motion seen as the **Pronation and Supination** movements. The axis is inclined 42° upwards from the transverse plane and 16° medially from the sagittal axis [12]. Thus the supination and pronation are considered as the coupled motion.

13.4.3 Kinematics at Subtalar Joint: Coupling Motion in Nonweight Bearing

As we have seen that the motion at the subtalar joint is a triplanar due to its orientation of the axis, the supination and pronation are coupled with adduction/abduction, inversion/eversion, and plantarflexion/dorsiflexion movements. An acronym has been suggested below to remember the subtalar joint coupled motion in nonweight bearing where the **calcaneum is moving** over the talus.

- **SADDIP—**Supination is coupled with **ADD**uction, **I**nversion, and **P**lantarflexion
- **PABED—**Pronation is coupled with **AB**duction, **E**version, and **D**orsiflexion.

13.4.4 Kinematics at Subtalar Joint: Coupling Motion in Weight Bearing

In weight-bearing subtalar kinematics, the key to remember is the motion of the talus over the calcaneum and refer to the acronym used above.The weight-bearing calcaneus will continue to contribute to the inversion/eversion component. The other two coupled components, i.e., abduction/adduction and dorsiflexion/plantarflexion, would be accomplished by movement of the talus in the opposite direction of the calcaneum.

Thus the supination in weight bearing would be coupled with abduction and plantar flexion of talus and inversion of calcaneum (SABID), whereas the pronation would include adduction and plantarflexion of the talus and eversion of the calcaneum (PADEP).

13.4.5 Close Chain Kinematics at Ankle Joint

Supination

1. Subtalar joint supination in a weight-bearing.

↓

2. The coupled component of talus abduction carries the mortise (the tibia and fibula) laterally.

↓

3. Produce lateral rotation of the leg.

Pronation

1. Subtalar joint pronation in a weight-bearing.

↓

2. The coupled component of talus adduction carries the mortise (the tibia and fibula) medially.

↓

3. Produce medial rotation of the leg.

Clinical Implication If a patient has flatfoot on weight bearing (pronated), the calcaneum would be everted, and the talus would be adducted and plantar flexed due to which the leg would rotate medial or medial tibial torsion could be seen. Compensatory close kinematic changes would also take place in the segments above.

Subtalar Neutral In weight bearing, the angle made by a line passing through mid of calf and mid calcaneum determine the subtalar neutral position. The subtalar neutral position is considered as deviation up to 2° calcaneal valgus [12].

13.5 Transverse Tarsal Joint

The transverse tarsal joint consists of Talonavicular (between talus and navicular medially)and Calcaneocuboid joint (between calcaneum and cuboid laterally).

13.5.1 Ligaments of Talonavicular Joint

- Medial Stability—**Deltoid ligament**
- Lateral Stability—**Bifurcate ligament**
- Inferior Stability—**Spring Ligament** (calcaneonavicular ligament)

13.5.2 Ligaments of Calcaneocuboid Joint

- Lateral Stability—Bifurcate ligament (also known as the **calcaneocuboid ligament)**

- Dorsal Stability—**Dorsal calcaneocuboid ligament**
- Inferiorly Stability—**plantar calcaneocuboid (short plantar) and the long plantar ligaments**

13.5.3 Transverse Tarsal Joint Kinematics

It functions as the transitional link between the hindfoot and the forefoot, serving to: [12]

1. Add to the supination/pronation range of the subtalar joint and.
2. Compensate the forefoot for hindfoot position—the tarsometatarsal joint would always try to maintain a good contact with the ground while compensation and thus go in Supination or Pronation twist as explained below.

13.5.3.1 Supination Twist

The supination twist is a kinematic function of the transverse tarsal joint seen when the subtalar joint goes into full pronation in weight-bearing position. The calcaneum would go into eversion, whereas the talus would be adducted and plantarflexed such that the transverse tarsal joint would be doing supination twist to maintain the foot contact with the ground and counter the pronation at the subtalar joint. The supination twist occurs around the second ray axis (anatomical axis of the foot), where the inversion component is common from first to the fifth ray.

13.5.3.2 Pronation Twist

The pronation twist is a kinematic function of the transverse tarsal joint seen when the subtalar joint goes into full supination in weight-bearing position. The calcaneum would go into inversion, whereas the talus would be abducted and dorsiflexed such that the transverse tarsal joint would be doing pronation twists to maintain the foot contact with the ground and counter the supination at the subtalar joint. The pronation twist occurs around the second ray axis (anatomical axis of the foot), where the eversion component is common from the first to fifth ray.

13.6 Metatarsophalangeal Joints (MTP)

The MTP are condyloid synovial joints with two degrees of freedom, including extension/flexion and abduction/adduction [12]. The joint is stabilized by joint **capsule**, **plantar plates, collateral ligaments, and the deep, transverse metatarsal ligament** [12]. During the end of the stance phase of walking, toes extension or dorsiflexion at metatarsophalangeal joints helps in the ground clearance of the foot. With the help of intrinsic foot and extrinsic muscles, metatarsal heads and toes help to balance the superimposed body weight.

13.6.1 Metatarsophalangeal Extension and the Metatarsal Break

The metatarsal break derives its name from the hinge or "break" that occurs at the MTP joints as the heel rises and the metatarsal heads along with toes remain weight bearing [12].

13.6.2 Windlass Mechanism

The integral movements to human locomotion include walking and running. The ability of the foot to manipulate with different surfaces at various ranges of speed is enhanced or facilitated by the interplay of joint movements within the small articulations of the feet. The foot structures adapt itself to different surfaces or terrain. This allows the foot complex to act as a compliant structure, thereby facilitating energy absorption and transfer of body weight. However, during the forward propulsion phase of the gait cycle, the foot performs as a stiff lever. The important structure which is responsible for transition and modulation for maintenance of the stiffness of the foot is the medial longitudinal arch. The windlass mechanism is depicted on each side by the dichotomous compliant- rigid characteristics of the foot. The arch behaves as a compliant and spring-like structure that flattens by compression in height and lengthens while in the early and midstance period of gait cycle. The longitudinal arch then recoils during the late stance phase of the gait cycle, meanwhile rising and shortening of the arch occurs, which may assist in the rigidity of the foot and may aid propulsion of the foot during push-off phase. Studies put forward that the arch changeover from compliant to the rigid function of the windlass action during the late stance phase with the help of the plantar fascia. According to Hick's windlass mechanism, dorsiflexion action at the metatarsophalangeal joint during the late stance phase lead to a winding effect at the plantar fascia around the metatarsal head. This pulls the plantar fascia causing the winding effect on the calcaneal bone.

13.7 Kinetics of Ankle Joint Complex

13.7.1 Muscles and Tendons

Plantarflexors

The main plantarflexors are the **Gastrocnemius and Soleus**. These muscles generate full strength for the plantarflexion at the ankle joint. The Gastrocnemius has two heads. Each head originates at each femoral condyle. The Soleus muscle originates from the upper tibia and fibula. Both these muscles meet at the Achilles tendon and insert at the posterior calcaneal surface. The other muscles of plantar flexion are the **tibialis posterior, flexor digitorum longus, flexor digitorum peronei, and flexor hallucis longus**. They can also act as plantar flexors, mainly in nonweight-bearing positions.

Dorsiflexors

The three muscles which produce dorsiflexion are the **tibialis anterior, extensor digitorum longus** and **extensor hallucis longus**. The extensor hallucis longus and tibialis anterior lie medial to the side of the axial rotation so they can also act as a supinator of the foot. The extensor digitorum longus, pass through the side or lateral aspect of the axis and cause pronation. The tibialis anterior muscle is a strong dorsiflexor following adduction and supination of the foot.

Invertors (Adduction–Supination)

The invertors of the foot are **tibialis posterior and tibialis anterior** as they lie medial to the axis. The tibialis posterior starts from the posterior portion of the tibia and fibula; in the course, it runs behind to the medial tibial malleoli and ends at the medial as well as the plantar aspect of the tarsal and metatarsals. The tibialis anterior originates from the lateral half of the upper aspect of the tibia and intersects the dorsal aspect of the foot, and is inserted at the medial portion of the medial cuneiform bone.

Evertors (Abduction-Pronation)

The evertor muscles tendons run along lateral to the axis of the foot. The evertors of the foot are **extensor digitorum longus, peroneus longus, tertius, and brevis.** The strongest evertors are peronei longus and brevis.

Intrinsic Foot Muscles

The sole of the human foot contains several muscles which originate and inserts within the plantar surface of the foot and are known as intrinsic muscles. Most of these intrinsic muscles originate from the calcaneal bone and the inferior portion of the posterior segment of the foot, and they insert at or near the phalangeal bones. The intrinsic muscles are the keystone in supporting the foot arches. They help in assisting the extrinsic muscles as well. The great toe muscles (abductor hallucis, adductor hallucis, flexor hallucis brevis) end at the lateral aspect of the base of the phalanx and then into the two sesamoid bones, which articulates with the head of the first metatarsal. There are four muscles that originate from the sides of the adjacent metatarsal bone and are known as dorsal interossei. These muscle bellies fill the intermetatarsal gaps. The tendons of these dorsal interossei insert to proximal phalanges at its base and to the dorsal digital expansions of the foot. The first dorsal interosseus tendon attaches to the second toe medially, and the other three tendons pass laterally to the second, third, and fourth toes. The action of intrinsic foot muscle is to move the toes away from each other or to abduct from the axis of the foot. These muscles also aid in toe flexion and extension movements.

13.7.2 The Plantar Fascia

This is the structure that plays a crucial role in maintaining the longitudinal arches of the foot. This is an aponeurosis fibrous structure that has got many layers. This fascia originates from the medial tuberosity of the calcaneal bone and runs anteriorly, and splits itself into five fibrous bands. Each splits at the corresponding metatarsophalangeal joint level and allows passage of both short and long plantarflexor tendons. During walking, when metatarsophalangeal joints go into dorsiflexion, the plantar fascia becomes tight and the longitudinal arch raise.

13.7.3 The Heel Pad

The heel pad acts as a shock absorber to attenuate the peak plantar pressure during walking and running. The thick skin is connected to that of the periosteum of the calcaneus by some big vertical fibrous septa. During compression (walking and running) these chambers are distorted and contribute to the shock absorption along with the deformation of chambers and with the flow of the enclosed adipose tissue.

13.7.4 Arches of the Foot

The foot has three arches: two longitudinal—medial and lateral arches and one transverse arch (Fig. 13.2). The arches of the foot are formed by the tarsal and metatarsal bones, and which are supported by tendons and ligamentous structures. The shape of the foot arches behaves in the same way as a spring. This helps in weight-bearing function of the foot and acts as shock absorption during locomotion. It also provides flexibility to the foot by facilitating the arches during walking and running [13].

Medial Longitudinal Arch (MLA)

The medial arch is attached proximal at the heels and distally at the medial three metatarsophalangeal joint structures. The bony structures which are involved in the formation of the medial arch are the calcaneus, talus, navicular, three cuneiforms and the first three metatarsal bones. The medial arch contains two pillars. The anterior pillar is made up of medial three metatarsal bones, while the posterior pillar is constituted of tuberosity of calcaneal bone.

The muscular structures supporting the medial longitudinal arch are flexor hallucis longus, flexor digitorum longus, abductor digitorum brevis, tibialis posterior. This tibialis posterior muscle tendon has a vital role in maintaining the medial longitudinal arch as the damage to which can lead to a collapse in the arch. More than the bony structures, the ligaments provide more stability in maintaining the arches. The plantar aponeurosis helps in supporting the beam between the two pillars. The head of the talus is supported by the spring ligament. The stability of the medial arch is further obtained by the talocalcaneal and deltoid ligament [14]. For the

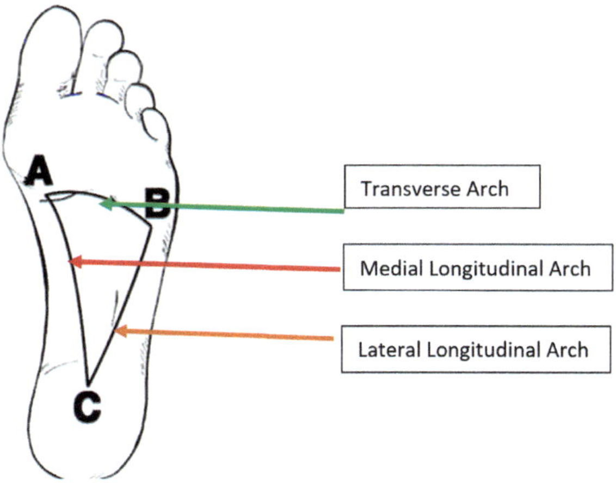

Fig. 13.2 Depicts longitudinal and transverse foot arch

biomechanical analysis, the foot could be considered as flat or high arch based on the normal length and height of the Medial Longitudinal Arch (MLA). The MLA height calculation could be done using the Chippaux Smirak Index (CSI), which is widely used for determining the arch of the foot as low or high [15]. The calculation could be taken in the following steps using Fig. 13.3 below.

- Step 1. Draw a straight line from point C to D
- Step 2. Draw the maximum wider segment of the metatarsal (Point A)
- Step 3. Draw the minimum wider segment (Point B)
- Step 4. Divide the length of the minimum wider segment by the maximum wider segment (B/A)

Interpretation: The higher index represents the flat MLA (low arch) and vice versa for the high arch. The type of foot based on the MLA arch classification was also given by Cavanagh and Rodgers [16] using the proportion of one third of the footprint by total surface area.

Lateral Arch

The lateral arch of the foot is less prominent and flatter compared to the medial longitudinal arch. The lateral arch touches the ground during quiet standing. It is made up of the calcaneus, cuboid bones along with the respective metatarsal bones (fourth and fifth metatarsal bones). The fourth and fifth metatarsal heads form the anterior pillar; calcaneal bone forms the posterior pillar. The plantar aponeurosis, long and short plantar ligaments provides stability for the lateral arch. The plantar ligaments behave like bowstrings underneath the arch [13].

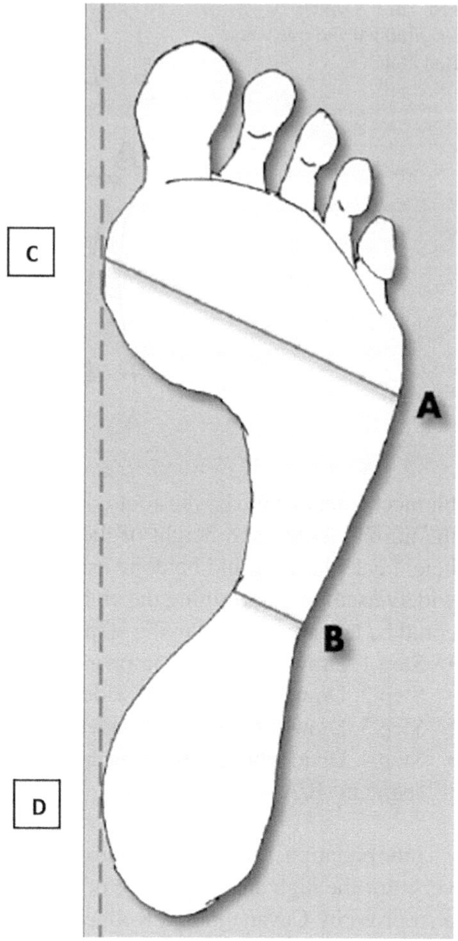

Fig. 13.3 Diagrammatic representation for calculation of MLA height using Chippaux-Smirak Index

Transverse Arch

The transverse arch is situated in the coronal plane and consists of bases of metatarsals, the cuboid and cuneiform bones. This arch is maintained by wedge-shaped intermediate and lateral cuneiform bones. Both medial and lateral arches of the foot behave as a pillar to the transverse arch. The arch curvature is maintained by tendons of the fibularis longus and tibialis posterior. Both the tendons of these muscles cross the sole of the foot. The adductor longus, in assistance with the fibularis longus tendon along with deep, transverse ligaments, stabilize the transverse arch of the foot [13].

13.8 First Ray Biomechanics

The first ray is a sole supporting unit at the distal end segment of a closely packed medial longitudinal arch. This structure allows the weight bearing of the foot during both stance and ambulation to withstand ground reactive forces. It is made up of the first metatarsal bone along with inner cuneiform. The placement of this articulation is very important as it bisects both the transverse and medial longitudinal arches. The first metatarsal bone and the first ray is the shortest but are the strongest. In the forefoot, it is one of the key weights bearing points. During the propulsive phase of the gait cycle, the shortest metatarsal bone moves into plantarflexion to maintain the connection with the underlying supporting surface. The first metatarsal inclines up to 15–25°. These degrees of freedom are the highest of all the metatarsals. The first metatarsal takes up to 40% of body weight during the stance phase. Ligamentous structures covering the joint provide stability at the first metatarsal-cuneiform joint articulation. The anterior and posterior tibia tendons along with the ligamentous structures provide support and stability at the first ray. The functional stability of the first ray is dependent on co-contraction of agonist-antagonist muscle [17].

13.9 Pathomechanics of the Ankle and the Foot

Hallux Valgus/Bunion

Hallux valgus is one of the most common foot deformities consisting of lateral deviation of the hallux at the metatarsophalangeal joint. In context to the foot axis (second toe), there is medial deviation of the first metatarsal and pronation at the hallux. Hallux valgus may be caused either due to extrinsic or intrinsic factors. The extrinsic factors include footwear like high heels which have a narrow toe box and allow limited toe mobility. Patients reported having collagen disorders, rounded shape of the MTP joint surfaces, the relatively long first ray of the foot, absence or amputation of the second toe, and hypermobility at the tarsometatarsal joint. The deviation, with the injury to the muscles, can cause displacement or subluxation of the sesamoids. This leads further imbalance of forces between the muscles and tendons. The flexor, extensor and adductor of the hallucis thus impose varus moment on the metatarsophalangeal joint. The abductor would go into adduction that impairs the counterbalance on the adductor.

Hallux Rigidus

Hallux rigidus or hallux rigidus is a degenerative process involving the first metatarsophalangeal (MTP) joint. Motion is restricted by periarticular osteophytes, especially in dorsiflexion. This condition is mostly seen in adults. Hallux rigidus can occur as a result of acute trauma like a fracture or forced hyperextension, or plantar flexion. It is most commonly seen due to the microtrauma to the articular cartilage overtime, which occurs repeatedly. Hallux rigidus has been graded based on radiographic findings [18].

- Grade I: Mild hallux rigidus—joint space maintained with minimal osteophytes
- Grade II: Moderate hallux rigidus—some joint space narrowing and dorsolateral osteophytes
- Grade III: Severe hallux rigidus—significant joint space narrowing and extensive osteophytes

Pes Cavus

The foot is supinated or high arch when the vertical height of the medial longitudinal arch is increased.

Pes Planus

The foot is pronated or flat foot when the medial longitudinal arch is collapsed.

Metatarsalgia

"Pain and inflammation in the forefoot region, involving the metatarsals or associated joints is known as metatarsalgia" [19]. It occurs due to the repetitive loading of the forefoot during gait cycle, especially at the midstance (from heel rise to toe off) phase. There are two types of metatarsalgia. Propulsive metatarsalgia is caused by stress during the propulsive phase of the gait cycle. Nonpropulsive metatarsalgia occurs in any phase other than the propulsion, including the swing phase [19].

Plantar Fascitis

"Plantar fasciitis is a common musculoskeletal disorder characterised by pain involving the inferomedial aspect of the heel that is exacerbated following periods of non-weightbearing" [20]. The diagnosis of plantar fasciitis is mostly based on clinical criteria. Pain is centralized to the medial tubercle of the calcaneal bone. On ultrasound, there will be diffuse or localized hypoechoic areas within a thickened calcaneal attachment. This condition is thought to be purely mechanical in origin. The etiology of plantar fasciitis is multifactorial which could be due to mechanical overload, overuse injury, the onset of lower limb musculoskeletal injury, pes planus, and subtalar joint pronation. Ground reaction and transient impulsive forces are linked to increase internal compressive stress within the heel during the gait cycle. An excessive tensile strain within the plantar fascia could increase the risk of plantar fasciitis. Tensile stress like nonuniform loading of the plantar fascia, bending force, shear, and compression force within the fascia can also cause exacerbation of the inflammation leading to plantar fasciitis. Age-related degenerative changes are also considered to cause this condition [20].

Ankle Sprain or Lateral Ligament Sprain

Injuries to the lateral ligaments of the ankle complex are among the most common injuries. Prevalence of the ankle sprain is common in both men and women. The most common predisposition for a lateral ankle sprain is the history of at least one previous ankle sprain. In sports like basketball, recurrence rates are as high as 70% [21]. Repetitive sprains have also been linked to high risk of development osteoarthritis and ankle articular degeneration. There can be mechanical or functional instability among the patients with a repetitive ankle sprain. Pathologic laxity after ankle-ligament injury is the mechanical cause. Functional instability is due to the occurrence of recurrent ankle instability and the patient having the sensation of joint instability due to poor proprioception and neuromuscular deficits.

Lateral ankle sprains most commonly occur due to excessive supination at the hindfoot about an externally rotated lower leg immediately after the initial contact of the hindfoot during gait cycle or landing from a jump. Excessive inversion and internal rotation at the hindfoot, along with the external rotation of the lower leg, results in lateral ligament strain. Increased plantar flexion during the early heel contact appears to increase the risk for the development of lateral ankle sprain. The anterior talofibular ligament is the first to undergo the injury, followed by the calcaneofibular ligament. Injury to the PTFL is typical only in severe ankle sprains and is often accompanied by fractures or dislocations, or both [21].

13.10 Summary

The ankle joint complex is the foundation for human body stability and mobility. The presence of various muscles and ligaments makes the structure unique and complex. In the close kinematics chain, any changes at the foot would lead to adaptive changes in the proximal segments.

References

1. Alazzawi S, Sukeik M, King D, Vemulapalli K. Foot and ankle history and clinical examination: a guide to everyday practice. World J Orthop. 2017;8(1):21.
2. Hessert MJ, Vyas M, Leach J, Hu K, Lipsitz LA, Novak V. Foot pressure distribution during walking in young and old adults. BMC Geriatr. 2005;5(1):1–8.
3. Tudor-Locke C, Craig CL, Aoyagi Y, Bell RC, Croteau KA, De Bourdeaudhuij I, Ewald B, Gardner AW, Hatano Y, Lutes LD, Matsudo SM. How many steps/day are enough? For older adults and special populations. Int J Behav Nutr Phys Act. 2011;8(1):1–19.
4. Cutts S, Obi N, Pasapula C, Chan W. Plantar fasciitis. Ann R Coll Surg Engl. 2012;94(8):539–42.
5. Brockett CL, Chapman GJ. Biomechanics of the ankle. Orthop Trauma. 2016;30(3):232–8.
6. Tan BH, Nather A, David V. Biomechanics of the foot. In Diabetic foot problems; 2008. p. 67–75.
7. Delco ML, Kennedy JG, Bonassar LJ, Fortier LA. Post-traumatic osteoarthritis of the ankle: a distinct clinical entity requiring new research approaches. J Orthop Res. 2017;35(3):440–53.
8. Holowka NB, Lieberman DE. Rethinking the evolution of the human foot: insights from experimental research. J Exp Biol. 2018;221(17).
9. Ghanem I, Massaad A, Assi A, Rizkallah M, Bizdikian AJ, El Abiad R, Seringe R, Mosca V, Wicart P. Understanding the foot's functional anatomy in physiological and pathological conditions: the calcaneopedal unit concept. J Child Orthop. 2019;13(2):134–46.
10. Hashimoto T, Sakuraba K. Strength training for the intrinsic flexor muscles of the foot: effects on muscle strength, the foot arch, and dynamic parameters before and after the training. J Phys Ther Sci. 2014;26(3):373–6.
11. Hernández-Díaz C, Saavedra MÁ, Navarro-Zarza JE, Canoso JJ, Villasenor-Ovies P, Vargas A, Kalish RA. Clinical anatomy of the ankle and foot. Reumatologia Clinica. 2012;8:46–52.
12. Levangie PK and Norkin CC. Joint structure and function: a comprehensive analysis. 2011.
13. Gwani AS, Asari MA, Ismail ZM. How the three arches of the foot intercorrelate. Folia Morphol. 2017;76(4):682–8.
14. Abboud RJ. (i) Relevant foot biomechanics. Curr Orthop. 2002;16(3):165–79.
15. Queen RM, Mall NA, Hardaker WM, Nunley JA. Describing the medial longitudinal arch using footprint indices and a clinical grading system. Foot Ankle Int. 2007;28(4):456–62.

16. Cavanagh PR, Rodgers MM. The arch index: a useful measure from footprints. J Biomech. 1987;20(5):547–51.
17. Glasoe WM, Yack HJ, Saltzman CL. Anatomy and biomechanics of the first ray. Phys Ther. 1999;79(9):854–9.
18. Chan O, Sakellariou A. Hallux rigidus: a review. Orthop Trauma. 2020;34(1):23–9.
19. Besse JL. Metatarsalgia. Orthop Traumatol Surg Res. 2017;103(1):S29–39.
20. Schwartz EN, Su J. Plantar fasciitis: a concise review. Permanente J. 2014;18(1):e105.
21. Vuurberg G. Management of lateral ankle injuries: a multidimensional approach. Universiteit van Amsterdam; 2019.

Kinematics and Kinetics of Gait

14

14.1 Introduction

Locomotion is the inherent capacity of all animals including humans which requires a complex biomechanics of body segments to accomplish a sequence of dynamic activity commonly referred to as Gait. The concepts and principles of biomechanics are applied to understand the characteristics of human gait with a simplified model and assumptions. The translatory and rotatory motion at each body segments work in a sequence of coordinated activity driven by the neuromuscular system. In this chapter, we shall focus on gait cycle, determinants of gait, kinematic and kinetics of gait, and pathomechanics.

14.2 Phases of Gait Cycle

For easy understanding and analysis, the gait is divided into phases and sub-phase which constitute the entire gait cycle. Broadly the gait cycle constitutes of **Stance phase (weight bearing phase) and Swing phase (nonweight bearing phase).** A single gait cycle is described as completion of successive events of one lower limb in stance phase. For instance, the heel touch to ground of one foot for the first time and heel touch to the second time of the same lower extremity would constitute a single gait cycle. It has been proposed that the Stance Phase constitutes about 60% of the gait cycle and 40% is contributed by the Swing Phase [1, 2]. In the stance phase of gait cycle, there is an instance when only one foot has contact with the ground and it is known as the **period of single limb support** constituting 80% of the stance phase gait cycle. At the other instance called **period of double limb support**, both foot have contact with the ground and constitute 20% of stance phase gait cycle. The traditional terminologies of gait are simple to understand and include sub-phases such as heel strike, foot flat, midstance, heel off, toe off, acceleration, midswing, and deceleration. Further analysis and phase classification have been

A. Hazari et al., *Conceptual Biomechanics and Kinesiology*,
https://doi.org/10.1007/978-981-16-4991-2_14

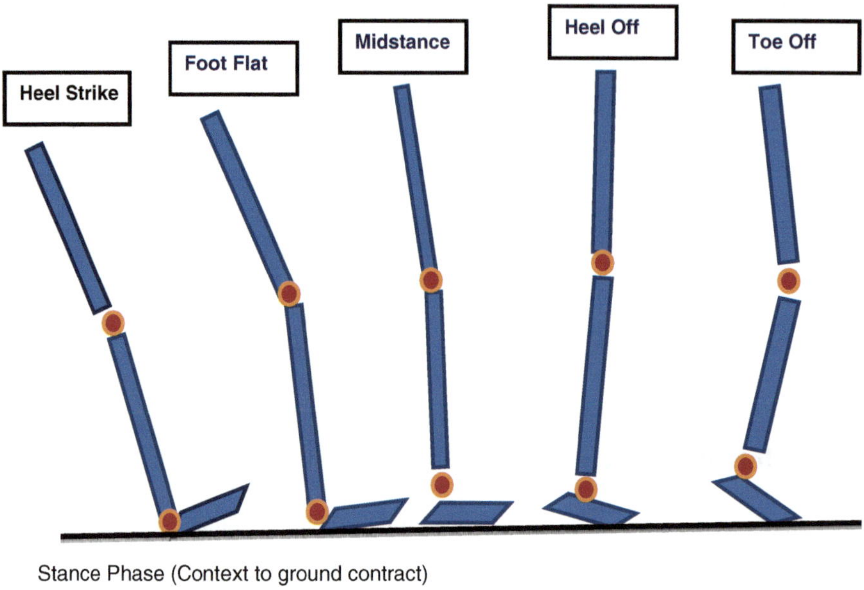

Fig. 14.1 Depicts sub-phases of stance phase in single gait cycle (traditional terminologies)

given by Rancho Los Amigos [3]. In this chapter we shall use traditional terminologies as shown in Fig. 14.1 for stance phase and Fig. 14.2 for swing phase, respectively.

14.3 Kinematic and Kinetic Analysis of Gait-2D

The 2D gait analysis is used commonly in clinical settings for patient population. The common devices used are static and dynamic foot scanner which yields the kinematic and kinetic spatiotemporal parameters of the gait. The spatial parameters consist of distance dependent variables, whereas the temporal parameters consist of time-dependent variables as explained below.

14.3.1 Spatial Kinematic Variables

Stride Length In one complete gait cycle, the stride length is measured as the straight distance from point of first heel strike (start of one gait cycle) to the point of second heel strike (start of second gait cycle) of the same lower limb.

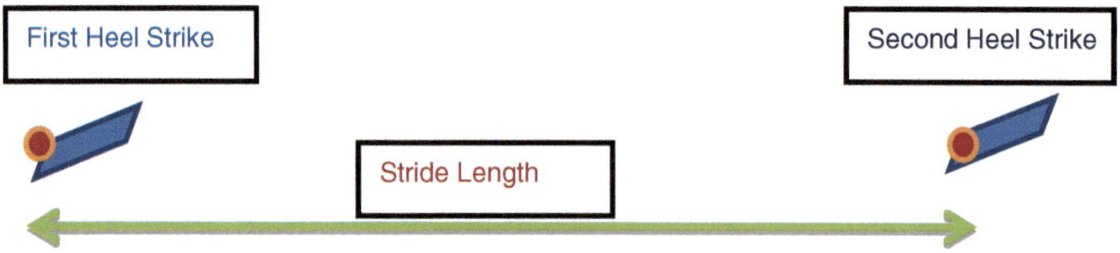

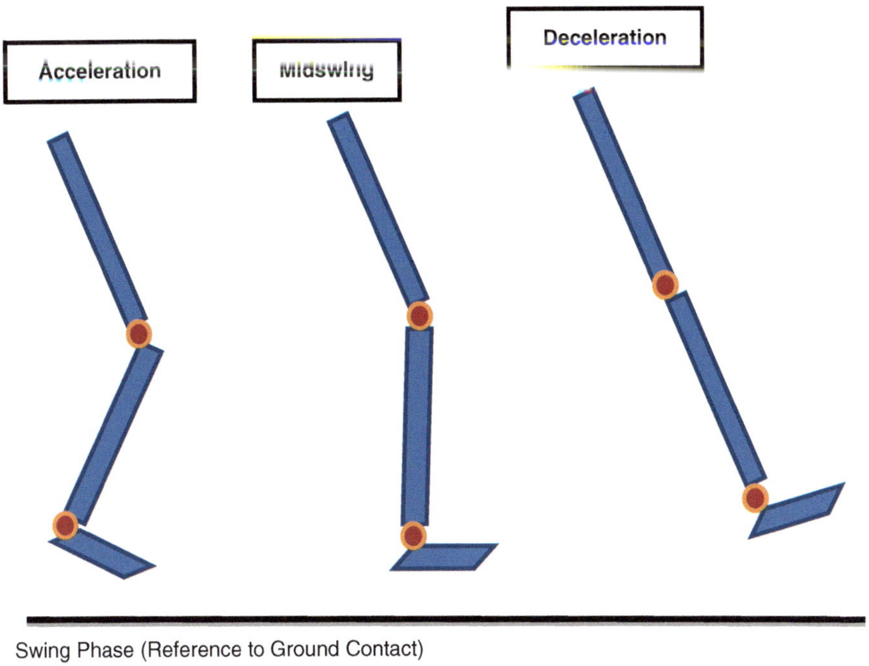

Fig. 14.2 Depicts sub-phases of swing phase in single gait cycle (traditional terminologies)

Note 1: The stride length can be calculated using any two successive events of the gait cycle. We have taken here heel strike as reference event. One can take foot flat or midstance or any other event as well. The only point to remember is use the same event for calculation.

Note 2: Using events from the stance phase has more reliability if done manually for clinical use as accurate point on ground contact. Swing phase measurements can be difficult. This applies to all spatiotemporal gait parameters.

Step Length In one complete gait cycle, the step length is measured as the straight distance from point of first heel strike one limb to the point of first heel strike of the other lower limb.

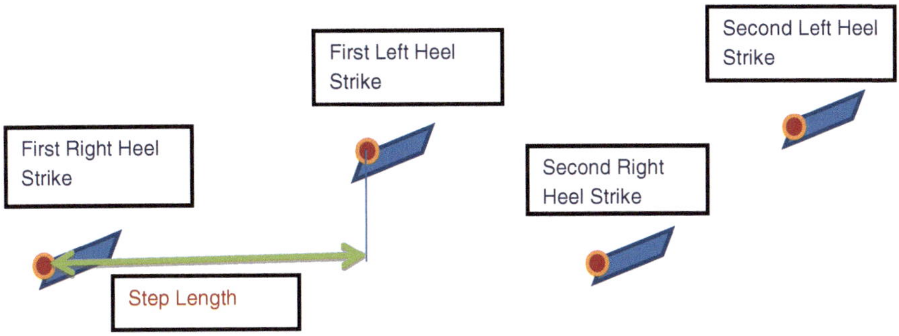

Note: The step length can be calculated using any two successive events of the gait cycle. We have taken here heel strike as reference event. One can take foot flat or midstance or any other event as well. The only point to remember is use the same event for calculation.

Step Width In one complete gait cycle, the step width is measured as the perpendicular distance from point of first heel strike one limb to the point of first heel strike of the other lower limb.

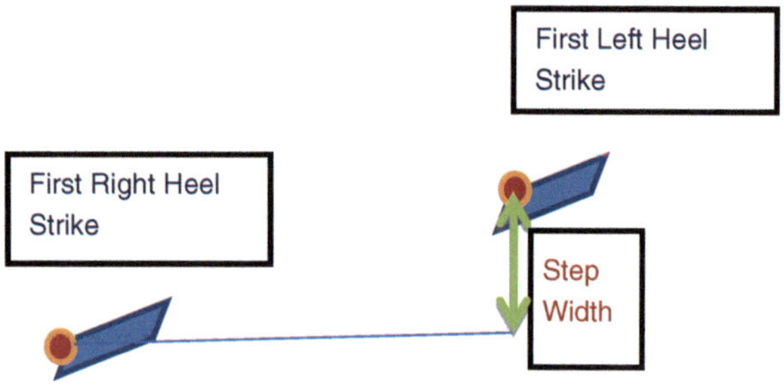

Degree of Toe-Out It is measured by the angle formed between the line of progression and the second toe (Fig. 14.3). The normal angle ranges from 7 to 10° [2, 4].

14.3.2 Temporal Kinematic Variables

Single-Support Time The duration in which the only one limb is in contact with the ground for one complete gait cycle.

Double-Support Time The duration in which both limbs are in contact with the ground for one complete gait cycle.

Stride Duration The time elapsed in completion of one stride length. It can also be referred as the gait cycle duration.

Step Duration The time elapsed for completion of one step.

Cadence Number of steps per unit of time measured as steps/min. Studies suggested 110 steps/min for men and 116 steps/women as typical cadence, respectively [2, 5]. It has been reported that shorter step length would increase the cadence for constant velocity [2, 6]. As cadence increases, period of double support decreases and running commences when cadence is above180 steps/min [2]. **Gait Speed:** represents the rate of progression and measured as distance/displacement covered per unit time.

The kinematic variable of interest in dynamic gait has been presented in Fig. 14.4.

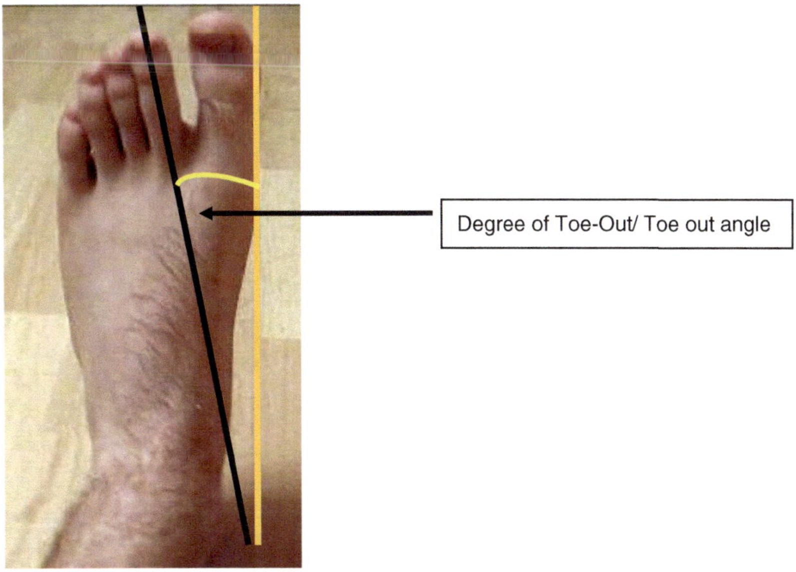

Fig. 14.3 Schematic illustration of toe-out angle formed between the line of progression and axis of the foot (second toe)

Résultats Podométriques

	1	2	3
Area	60.0	32.0	26.0
Average P.	2287.5	2411.9	1832.1
Maximal P.	4311.3	4311.3	4311.3

—— GaitLine —— MaxLine

Résultats Spatio-Temporels

Step	Duration	(ms)	400	890
Gait Cycle	Duration	(Ms)	920	1290
Single Stance	Duration	(ms)	500	
Double Stance	Duration	(ms)	90	
Swing	Duration	(ms)	990	1310
Stride	Duration	(ms)	1020	1880
Step	Length	(mm)	383	312
Gait cycle	Length	(mm)	765	695
Angle		(°)	8.97	3.28

- 18% 29% 36% 44% 51% 59% 67% 77% 84% 94%

Medicapteurs Software - France - USA

Fig. 14.4 Spatiotemporal Gait parameters

14.3.3 Spatial Kinetic Variables

Plantar Pressure Distribution The plantar pressure distribution is important kinetics to determine the sites of high pressure and overall pressure distribution. Plantar pressure is measured as force per unit area. The most commonly used variables for understanding the plantar pressure distribution include **average pressure** (total force/total contact surface area), **maximum plantar pressure** (peak force per unit area), and **forefoot to hind foot pressure ratio** (determined by the location of center of pressure). The plantar pressure distribution for static gait has been represented in Fig. 14.5.

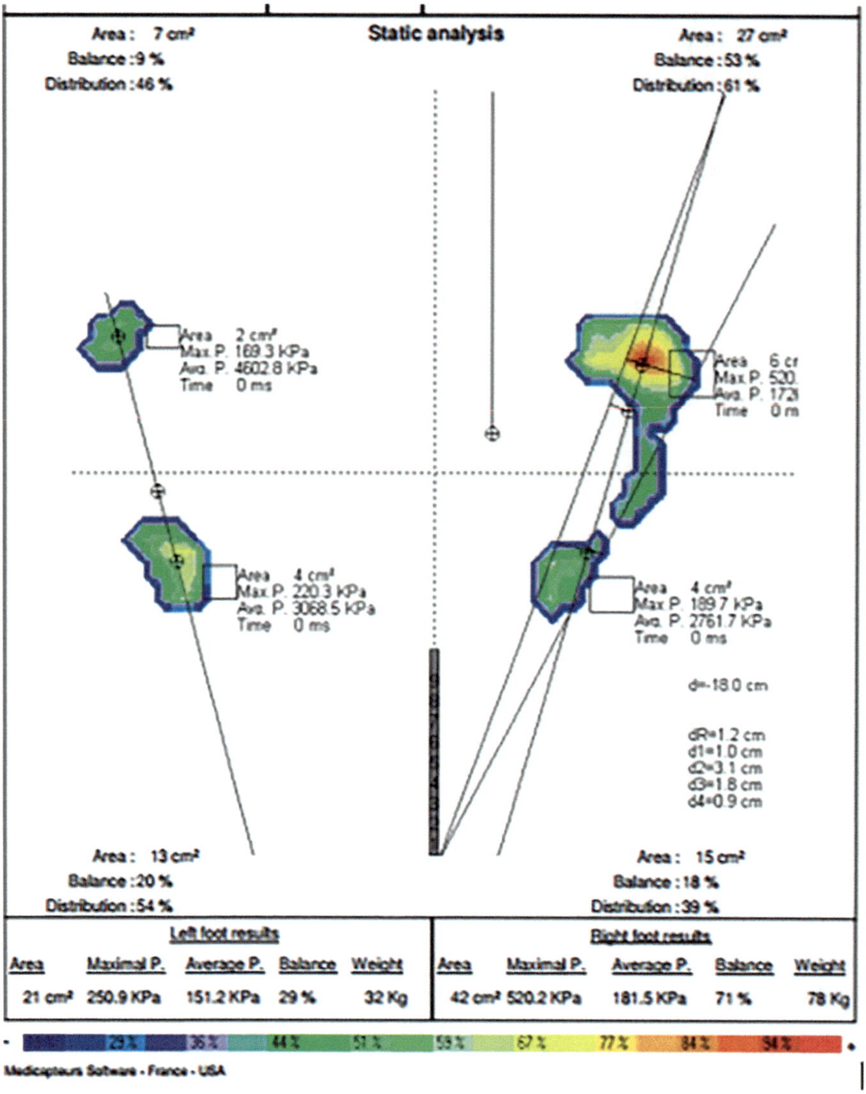

Fig. 14.5 Showing plantar pressure distribution for static gait analysis

14.3.4 Temporal Kinetic Variable

Pressure Time Integral The contact time during dynamic gait has high significance for the plantar pressure analysis. During the gait progression, the frictional force, shear force, and gravity combine to generate an impulsive force as shown in Fig. 14.6.

14.4 Kinematic and Kinetic Analysis of Gait-3D

The analysis of gait for research requires advanced instruments for accuracy. The motion analysis system with infrared cameras is often used in biomechanical lab to generate information for gait analysis using the inverse dynamics. The variables obtained are accurately quantified in all three planes as compared to 2D analysis. The use of retroreflective joint markers enhances the accurate application of 3D gait analysis as they need to be worn directly over the joint surface to avoid artifacts (Fig. 14.7). However the biomechanical gait analysis without markers has become popular in many research settings (Fig. 14.8).

14.4.1 3D Kinematic Variable

The kinematic variables of interest in 3D include joint angle, velocity, and acceleration as shown in Fig. 14.9.

Joint Angle: The angle formed at different segments of entire body can be obtained.

Joint Velocity: It is measured as the rate change of joint angle.

Joint Acceleration: It is measured as the rate change of joint velocity.

14.4.2 Kinetic Variable

In the previous chapter, we have learnt the importance of COG, COM, and LOG for posture and balance. Since human gait requires static and dynamic stability throughout, the progression of gait is very much dependent upon these variables. The kinetic

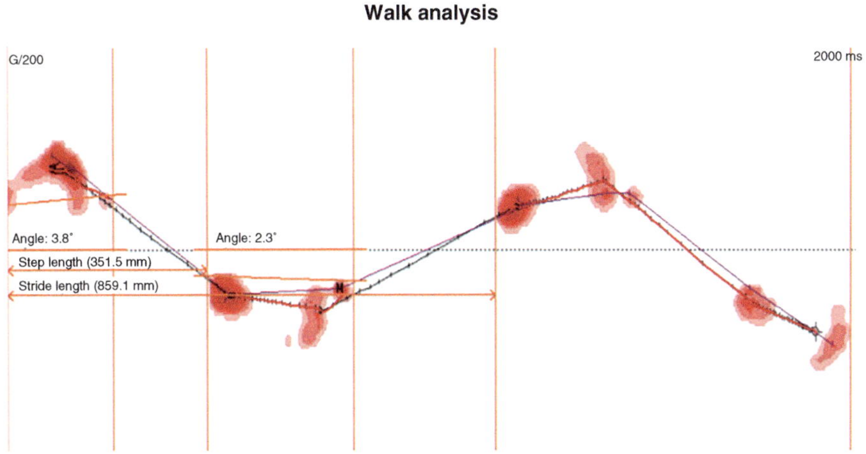

Fig. 14.6 Showing impulsive force (pressure × time)

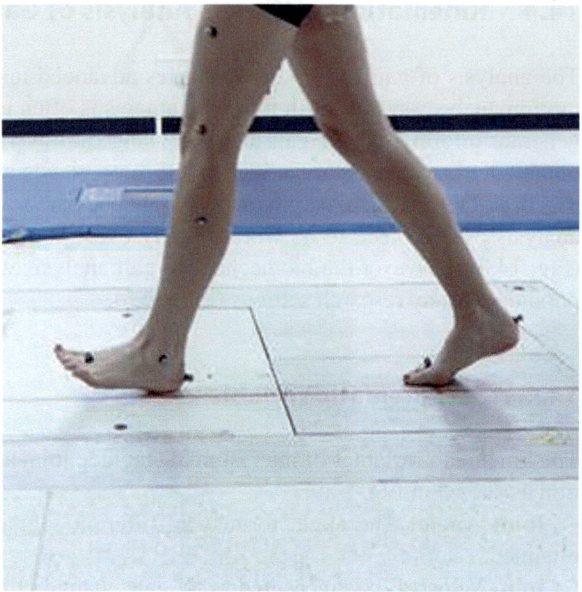

Fig. 14.7 Gait analysis with retroreflecting joint markers using infrared cameras and force platform (source—Jupiter Gait Lab, Thane, Maharashtra, India)

variables of interest in 3D include power, Center of Pressure (COP), external moment (GRF), and joint internal moment.

Power The power of a body is defined as the ability to do work. In context to human biomechanics, the power is related to muscular activity to generate or absorb the forces. When a muscle contracts, the power is generated and a positive work is done. During eccentric contraction power is absorbed and muscles perform negative work. The power generated or absorbed across a joint is the product of the net internal moment and the net angular velocity across the joint [2, 7]. The power is obtained using the inverse dynamic laws in the motion analysis systems.

Center of Pressure The COP for human gait produces a characteristic pattern starting at the posterolateral edge of the heel during heel strike and moving medially towards the metatarsal head of first and second digits as the gait progresses (Fig. 14.10).

Joint Moment It is evident to understand that in order to be stable during both static and dynamic gait; the body needs to be in equilibrium. In simple words, the external force created by the ground **(external moment)** should be balanced by the internal force of the muscle **(internal moment).** At every stage of gait cycle, different muscle act to create an internal joint moment to overcome the force created by the LOG and not allow the joint segments to collapse otherwise. We can understand this by analyzing our ability to walk with upright posture without falling down. We know that gravity pulls all the body towards the ground and in that case we would not be able to stand against the gravity and rather crawl for locomotion. If we are

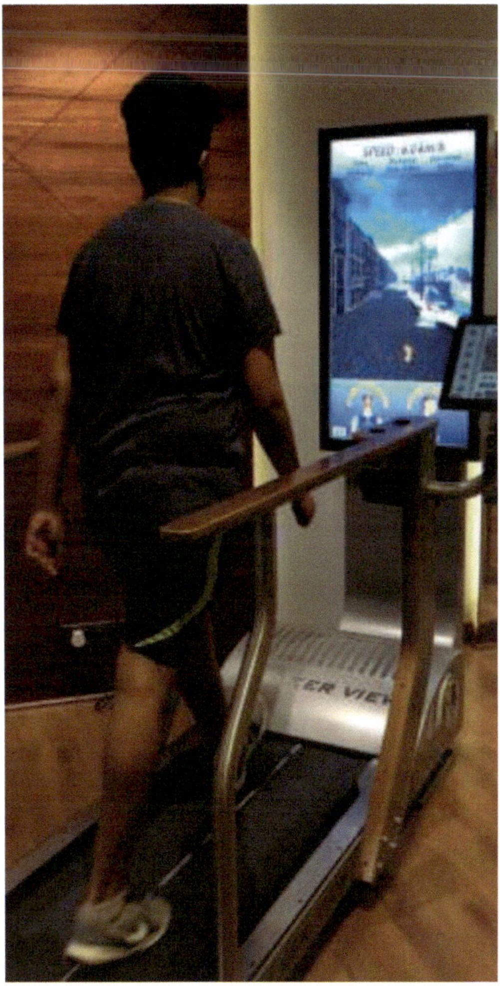

Fig. 14.8 Gait analysis without using joint markers (source—Abhinav Bindra Targeting Performance, Chandigarh, India)

able to walk, run, and perform activities against the force of gravity, it clearly suggests that our musculoskeletal system has overcome the gravitation force and muscle is acting without our knowledge at times by creating internal joint moments. It is also to be noted that LOG keeps shifting at different phases of gait cycle due to shift in COM and COP. Therefore, it is important to understand the net internal joint moments at each phase of gait cycle. Referring to Figures 14.31, 14.32 of the textbook "Joint Structure and Function" by Levangie and Norkin [2] and Fig. 14.11 here, let us analyze the external and internal joint moments at stance phase of gait cycle in sagittal plane.

General Principal As discussed, the internal joint moment is created by muscles against the external joint moment by ground reaction force. Thus is it first important to analyze the direction of moment created by the ground reaction force vector

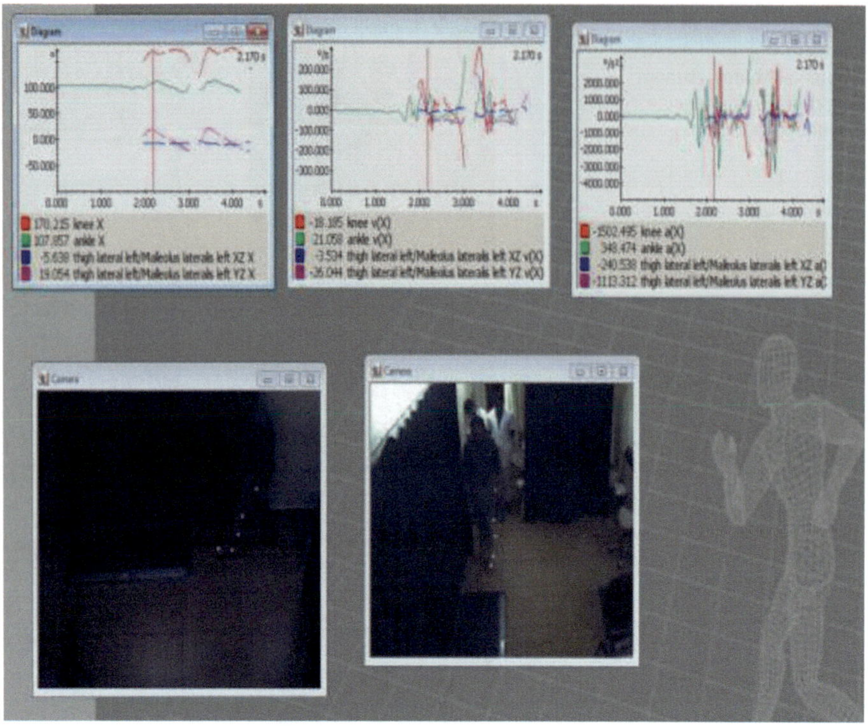

Fig. 14.9 Graphical representation of kinematic variables at ankle and knee using SIMI motion analysis system (angle in degree, velocity in degree/s, and acceleration in degree/s^2) (source—Diabetic Foot Lab, KMC, Manipal)

(GRFV) and the muscle would act in opposite direction. If the GRFV falls posterior or anterior to the joint centers (represented by dots at ankle, knee, and hip), move the proximal segment toward the GRFV to determine the direction of external moment. This would be applicable to all phases of gait cycle. In addition if muscle works concentrically to overcome the external moment, we can say positive power/work is done and if muscle works eccentrically in the context here, then negative power/work would be seen.

Joint Moment at Initial Contact: (PA)*

The GRFV falls posterior to ankle, anterior to knee, and anterior to Hip. Thus the proximal segment at ankle, i.e. tibia would move posteriorly towards the GRFV creating an external plantarflexion moment (the angle between the tibia and dorsum of foot would increase when tibia moves posterior as reciprocal to plantarflexion at ankle joint). Similarly, the proximal segment at knee, i.e. femur, would move anterior since GFRV is passing anteriorly and would create an external extension moment at knee and at hip external flexion moment would be created because the pelvis would tilt anteriorly towards the GRFV. Now, the muscles would contract to generate an internal moment thereby a dorsiflexion moment at ankle, flexion

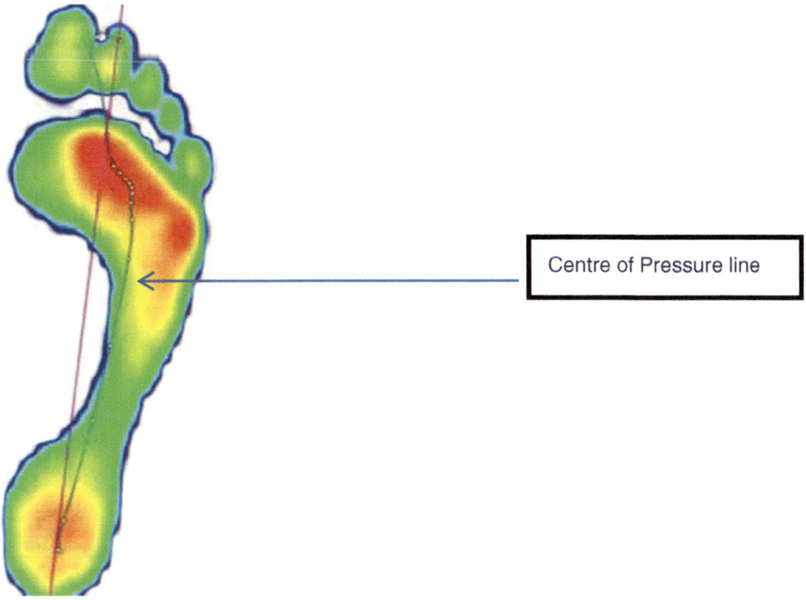

Fig. 14.10 Represents COP progressing during a gait cycle

moment at knee, and extension moment at hip would be taking place as shown in Fig. 14.11. In context to the muscle power, the tibialis anterior would contract concentrically to generate a positive power, hamstring would generate positive power by contracting concentrically at knee, whereas the quadriceps would contract eccentrically to absorb power and do negative work.

Joint Moment at Foot Flat: (PAK)*

In contrast to the initial contact phase, the GRFV at foot flat shifts posterior at knee joint and remains same at ankle and hip (Fig. 14.11). Thus the only difference in the internal joint moment would be extensor moment at knee joint where quadriceps would have to contract concentrically to produce knee extension and generate a negative power.

Joint Moment at Midstance: (PH)*

The GRFV at midstance is just opposite of foot flat. It shifts Anterior to Ankle, Anterior to Knee, and Posterior to Hip. Thus an internal plantarflexion moment at ankle, flexion moment at knee, and flexion moment at hip would be created as shown in Fig. 14.11.

Joint Moment at Heel Off: (PH)*

The GRFV at midstance is same as seen at midstance phase. Thus an internal plantarflexion moment at ankle, flexion moment at knee, and flexion moment at hip would be created as shown in Fig. 14.11.

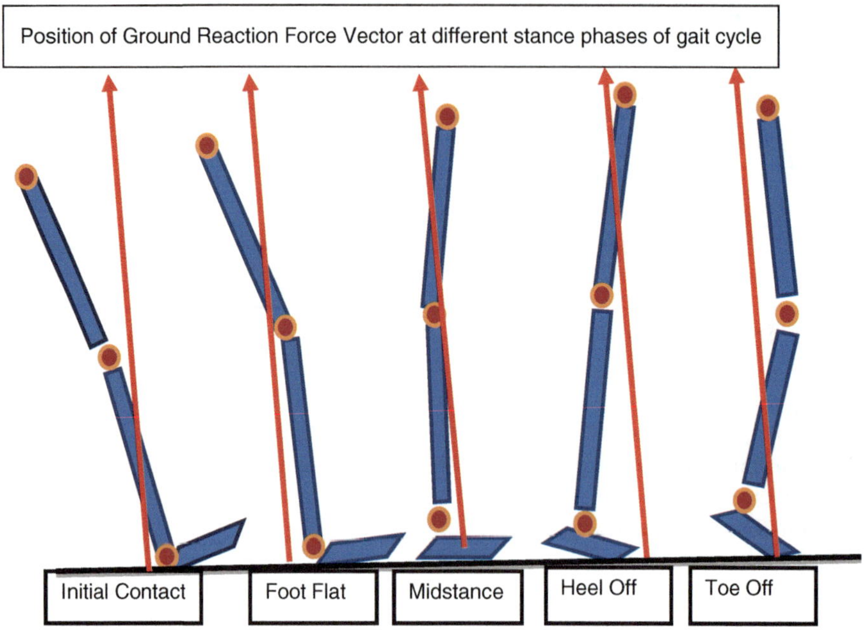

Fig. 14.11 Depicts the net internal moment by muscles against the external moment created by Ground Reaction Force (GRFV and COP) at stance phase of gait cycle

Joint Moment at Toe Off: (AA)*

The GRFV at toe off phase is seen anterior to ankle, posterior to knee, and posterior to hip. Thus an internal plantarflexion, extension and flexion moment at knee and hip respectively, would be generated by muscles of the respective joints.

Note 1: * The easiest method to remember the direction of GRFV at difference stance phase of gait cycle using traditional terminologies is to memorize the synonym given in asterisk where **second letter**,

- A = ankle
- K = knee
- H = hip

The direction of GRFV at given joints (A, K, H) is represented by the **first letter** (P = posterior or A = anterior). It is understood that if the direction for joint which is not mentioned would be opposite to what is mentioned. For example, GRFV at midstance and heel off is represented by **PH**. Thus the direction for GRFV at midstance and heel off is posterior (P), letter for direction to Hip (H), letter for joint. Since ankle and knee are not mentioned thus they would just be opposite to hip, i.e. anterior at knee and anterior at ankle. This is applied to all phases above. Try yourself for initial contact and foot flat, toe off.

Note 2: The GRFV is not applicable for swing phase as there is not contact with the ground. However muscles keep acting concentrically or eccentrically as per the required joint function. It does not require to work additionally against the external force created by ground reaction.

14.5 Close Chain Kinematics and Kinetics

In the above section, we have learnt about kinematics and kinetics of gait focused on the lower limb. In normal walking gait, the pelvis, upper extremity, and the trunk also play a significant role. The joint external and internal moments at the lower limb has a closed chain relationship with trunk, head, and upper extremity biomechanics. The noticeable motions are seen at pelvis, trunk, and arm in all three planes as described in Table 14.1. The muscle of the trunk (abdominals), spine, and shoulder actively participate to generate coordinated and smooth movements.

14.6 Biomechanics of Stair Gait

Stair gait is a common functional activity required in day to day activities. The stair gait is also an important part of physical rehabilitation. Thus it is important to focus on the biomechanics of stair gait in brief. Alike normal gait, the stair gait also consists of stance and swing phase contributing to 64 and 36% of total gait cycle [2]. During the stair ascent, the stance phase is subdivided into weight acceptance, pull up, and forward continuance, whereas the swing phase consists of foot clearance and foot placement as shown in Fig. 14.12 [2, 8].

Table 14.1 Closed kinematic relationship of lower limb with pelvis, trunk, and arm during normal walking

Plane	Pelvis	Trunk	Arm
Sagittal plane	Translates up and down in a sinusoidal wave like movement	Slight flexion and extension	Alternate forward and backward swing (if right foot is leading, right arm would swing backward and left arm would swing forward and vice versa)
Frontal plane	Hike on stance limb Drop on swing limb	Lateral leaning opposite to pelvic drop side	Slight adduction towards midline
Transverse plane	Forward or anticlockwise rotation on the leading foot	Backward rotation or clockwise rotation	Slight medial and lateral rotation

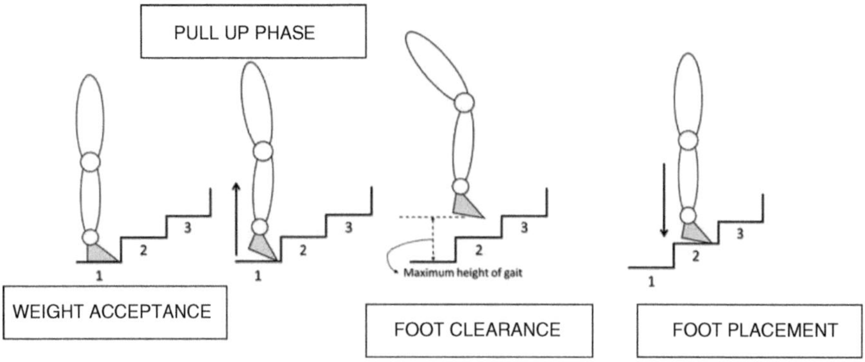

Fig. 14.12 Depict the phases of stair climbing

14.7 Pathomechanics

A number of factors can affect normal gait patterns. In the scope of the book, we shall focus on the musculoskeletal dysfunctions that can alter the gait biomechanics as discussed below.

14.7.1 Muscular Dysfunctions

Gluteus Medial Gait Weakness or paralysis of hip abductors causes altered pelvis kinematics and kinetics represented as lateral leaning during gait pattern (Trendelenburg gait).

Gluteus Maximus Gait Due to the paralysis of the gluteus maximus muscle, the internal extension moment at hip and pelvis is not possible against the external flexion moment. As the pelvis and hip need to balance itself, there is posterior thrust of the trunk during heel strike which is the characteristics of the gluteus maximus gait (lurching gait).

Calcaneal Gait Paralysis or weakness of the plantarflexors results in a calcaneal gait pattern characterized by short step length and excessive dorsiflexion, knee flexion of the affected side [2, 9].

Quadriceps Gait If the quadriceps gets paralyzed, the extension of knee would not be possible during gait, thus the individual manually pushes the knee into extension with support of arm which is known as the quadriceps gait or hand to knee gait. In case of patellar dysfunction also, the quadriceps gait may prevail.

14.7.2 Structural Dysfunction

Limb Length Discrepancy The significant difference in limb length can cause lever arm dysfunction and thus affect the normal gait pattern. The major compensation is seen at the level of pelvis. The shorter limp tends to drop pelvis excessively to maintain contact with the ground.

Altered Foot Arch The changes in the medial longitudinal arches of the foot as seen with pes planus and cavus can cause gait dysfunctions by causing kinematic and kinetic changes in the proximal segments as part of the close kinematic chain.

Altered Joint Alignments The joint alignments and angulations play an important role in determining the joint axis and their function. In case there is a change as seen with Q angle, femoral anteversion, tibial torsion, etc., the gait can be significantly affected as the body would try to compensate for the altered kinetics.

14.7.3 Neurological Impairments

The most commonly seen gait due to neurological impairments gait includes Parkinson's and cerebellar ataxia. The dysfunction of the higher mental function are responsible for such gait pathology.

14.7.4 Compensatory Gait

The pain at hip, knee, and ankle commonly creates a compensation to avoid pain by moving the GRFV closer to the joint so that muscles do not overwork and exacerbate the pain. This is commonly seen with hip joint pain, known as the antalgic gait.

14.8 Summary

The human gait is a very complexed and biomechanically well driven function of the neuromuscular systems. The coordinated muscle action allows the joints to perform kinematic and kinematics very efficiently. Any changes in the neuromuscular structure and function can lead to gait abnormalities.

References

1. Lamoreux LW. Kinematic measurements in the study of human walking. Bull Prosthet Res. 1971;10(15):3–84.
2. Levangie PK, Norkin CC. Joint structure and function: a comprehensive analysis. Philadelphia, PA: F.A. Davis; 2011.

3. Rancho Los Amigos National Rehabilitation Center. Pathokinesiology Service and Physical Therapy Department at Observational Gait Analysis. Downey, CA: Los Amigos Research and Education; 2001.
4. Murray MP. Gait as a total pattern of movement: including a bibliography on gait. Am J Phys Med Rehabil. 1967;46(1):290–333.
5. Rose J, Gamble JG, editors. Human walking, vol. 3. Baltimore, MD: Williams & Wilkins; 1994. p. 32.
6. Larsson LE, Odenrick P, Sandlund B, Weitz P, Oberg PA. The phases of the stride and their interaction in human gait. Scand J Rehabil Med. 1980;12(3):107–12.
7. Winter DA, Eng JJ, Ishac MG. A review of kinematic parameters in human walking. In: Craik RLB, Otais CA, editors. Gait analysis: theory and application. St. Louis, MO: Mosby; 1995.
8. Harper NG, Wilken JM, Neptune RR. Muscle function and coordination of stair ascent. J Biomech Eng. 2018;140(1).
9. Perry J. Kinesiology of lower extremity bracing. Clin Orthop Relat Res (1976-2007). 1974;102:18–31.

Biomechanics of Static and Dynamic Posture

15

15.1 Introduction

The postural control in human body is one of the most important biomechanical features which determine its quality of motion as well as the forces acting on the body. The erect posture like standing is required to allow the upper extremity to move through a wider range of motion while the lower body stabilizes the body. In contrast, the posture at running has to support mobility with lower limb and stability by the trunk and upper extremity. Thus a normal posture is very important to maintain the state of equilibrium in both static and dynamic phases. The state of equilibrium is identified as ability of the body to maintain the stability without the use of any external force or support. The human body is well designed to function and maintain the stability at rest as well as in motion which we will refer as the static and dynamic postural control in this chapter. The postural control for any physical body should be taken in context to the application of internal and external forces. The posture at rest where there is no major change in the primary position such as sitting and standing is known as the **static posture,** whereas posture at motion such as walking, running, jumping, etc., is known as **dynamic posture.**

15.1.1 Variables of Posture

The stability and postural control of a body is determined by its variables like base of support (BOS), Center of Mass (COM), Center of Gravity (COG), and Line of Gravity (LOG). Any disturbance in these variables would affect the postural control at rest or in motion which would require support from external means and we shall discuss in the upcoming section. Let us understand the variables of postural control in more detail here.

A. Hazari et al., *Conceptual Biomechanics and Kinesiology*,
https://doi.org/10.1007/978-981-16-4991-2_15

(a) **Base of Support (BOS):** The BOS for any physical body is determined as the rectangular area covered by its body parts touching the ground. For example, the BOS for human body would be the rectangular area covered by two feet as shown in Fig. 15.1. The larger the BOS, the better could be the postural control of a body and lesser would be the chances of fall (instability). For example, we feel more stable when we stand with feet apart in comparison to feet closer (Fig. 15.2a, b). In context to biomechanics, the BOS would control frontal and sagittal plane motions predominantly. Any moment within the BOS would be under the postural control and movement beyond the BOS would disturb the state of equilibrium and require external support. The BOS for a cow would be larger compared to human as the rectangular area would be larger covered under four feet instead of two as seen in human.

(b) **Center of Mass (COM):** The COM for a physical body is the point where the entire mass is assumed to be located. The COM is thus not fixed and changes with the postural changes. For example COM in sitting would be lower than in standing for all of us because the mass of the body has shifted down in sitting position. If an additional weight is added, the COM would shift towards the heavier part of the body. Ideally the COM for an adult human body is located at the vicinity of S2 vertebra. It should be noted that the closer the COM lies with BOS; the better would be the stability of the body. This is the reason it is easy to displace a taller object compared to shorter with the same BOS.

(c) **Center of Gravity (COG):** The COG for a body is the hypothetical point at which the force of gravity acts suggesting that COG can lie within (stable posture) as well as outside the body (unstable posture). In ideal posture the COG coincides with COM and lies at the point where the entire mass of the body is concentrated [1].

(d) **Line of Gravity (LOG):** The LOG can be seen as the force vector from the COG to the ground. In simple words, the line through which the gravitational force pulls act can be considered as the LOG. If the LOG lies within the base of support, the body would be stable and if lies outside it would be unstable (Fig. 15.3).

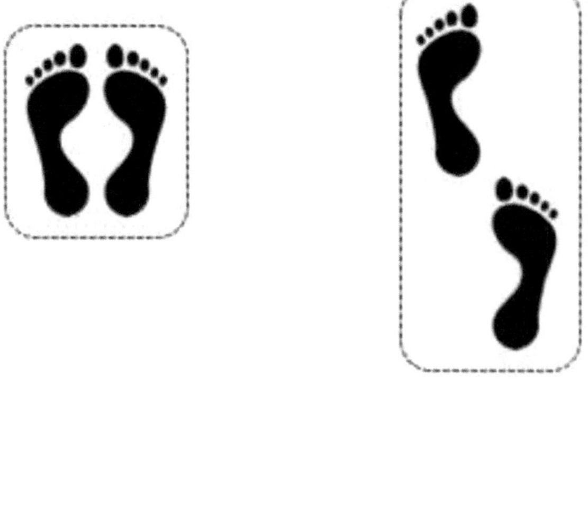

Fig. 15.1 Showing base of support at standing (static) and walking (dynamic). The rectangular area covered by the feet represents the BOS

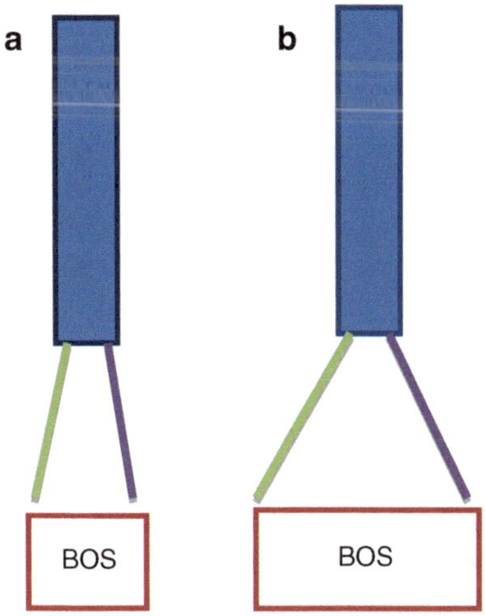

Fig. 15.2 (**a**) Smaller base of support when standing with feet close (**b**) larger base of support when standing feet apart

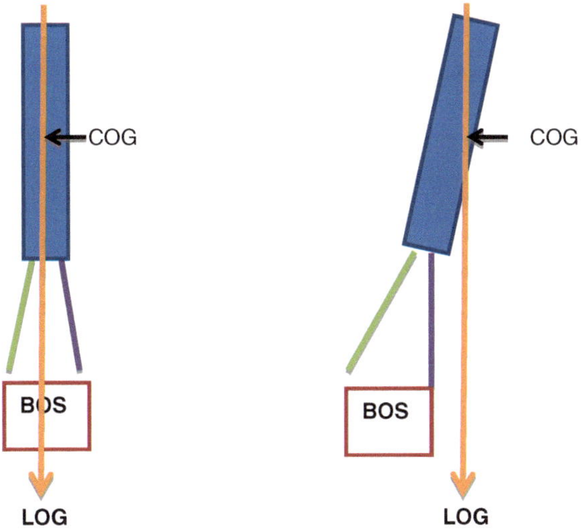

Fig. 15.3 Depicts relationship of LOG with BOS (body is stable when LOG lies within BOS (left) and unstable when outside BOS (right))

15.2 Mechanism of Postural Control

The posture is a function of balance and coordination. The balance and coordination in human body is determined by four major systems (1) Central Nervous System—the cerebellum plays the major part, (2) Vestibulocochlear system, (3) Proprioceptive feedback from the joint, and (4) Muscular contraction. Any dysfunction of either system can disturb the postural stability and balance. In the scope of this book and human biomechanics, we shall focus on postural control of the human body which is determined mainly by the interaction of the neuromuscular system. The brain sends signal to musculoskeletal systems to control the stability through active and passive structures. At rest, the passive structure is major contributors to posture and muscle play an important role in dynamic stability. As we have learnt that the application of external forces can disturb the equilibrium, the neuromuscular systems have the ability to response to maintain the postural control either by **reactive or proactive response**. The reactive response is seen when the body responds to a sudden change in the state of equilibrium as **compensatory mechanism.** Consider that you are walking in dark light and suddenly you step on a stone, the body would use its compensatory mechanism to control the posture. Similarly, you are standing and someone slightly pushes you without your understanding, you perturbate as the reactive response. In contrast, the proactive response comes through the **anticipatory mechanism** where we anticipate and respond accordingly. For example, you are walking and you know that you need to step on the stones to proceed ahead; the body will prepare itself for the consequent changes in prior and respond to maintain the postural control. The proactive response is also developed based on our past experiences.

In human biomechanics, the reactive and proactive responses to perturbation are controlled by muscle synergies or strategies [1, 2]. The reactive responses are controlled by muscle synergy, whereas the proactive responses are maintained by the body strategies.

Change in Support Strategies During instability, the first response is to step and increase the BOS which is known as the **stepping strategy** as seen with proactive response to a strong external force (someone forcefully pushes you). If the stepping strategy is inefficient to control the posture, holding an object for support would be required which is known as the **grasping strategy** as seen when we are in a state to fall. The change in support strategies is very efficient with larger perturbations [1, 2]

Head-Stabilizing Strategies The head stabilization strategies are used to keep the head over the body at all instances. The **head stabilization in space** is associated with vertical position of head in respect to environment. The vestibular and the cerebellar systems play an important role in this. During walking, the position of head is maintained with respect to the environment and held vertical over the COG. The **head stabilization on trunk** can be seen as the position of head with respect to trunk. During dance and forward leaning, the head follows the trunk and remains straight over it through inclined in respect to the horizontal plane. When the change in support strategies fail, the specific muscle activation helps to maintain the support which is known as the synergy [1]

Fixed-Support Synergies The specific pattern of activity and coordinated muscle action as a response to perturbation in standing posture is known as the muscle synergy [1].

Since the feet remains fixed within the BOS and the stability is regained against perturbation through muscle activation, it is known as fixed-support synergies. Two major synergies consist of ankle and hip synergy and they are the first response for postural control to perturbations or deviations in standing.

Ankle Synergy It can be considered as the compensatory response from contraction of muscles in distal to proximal pattern when we perturbate through sagittal plane motions. For instance, we are standing on a platform which is displaced suddenly anteriorly or posteriorly. If the platform is displaced posterior/backwards, LOG would fall anterior and we tend to fall forward. In response as ankle synergy is initiated where the posterior muscles would start contracting from distal to proximal. As the results of contraction of plantarflexors, soleus, knee and hip extensors the LOG would be maintained within the BOS to stabilize the body. If platform is displaced anteriorly, the anterior muscles from distal to proximal would start contraction.

Hip Synergy Unlike the ankle, the hip synergy is seen as contraction of muscles from proximal to distal pattern [1, 3]. For example, if we displace the platform forward, the hip synergies can be seen as sequential activation of abdominals, hip flexors, knee extensor, and ankle dorsiflexors.

15.3 Kinematics of Posture

The kinematics of posture is determined by the LOG passing through the joint centers. In clinical practice, the sagittal plane is used to determine the position of LOG using the plumb line. If the plumb line passes normally to joint centers as discussed below, the posture would be considered as an optimal posture. We shall learn about the optimal posture in standing and sitting as they are fundamental postures from where any other posture may be derived and commonly referred to as optimal posture. The deviation of the LOG will lead to abnormal biomechanics.

15.3.1 Sagittal Plane Analysis of Standing Posture

At standing, the LOG at different joints passes as described in Table 15.1. The easy method of remembering is to analyze from the top to bottom as alternative anterior and posterior where ankle is exception with LOG anteriorly placed. Now the kinematics of these joint centers can be determined by just moving the joint towards the LOG. In simple words, if LOG is posterior to joint center, the joint will move posterior and anterior if LOG lies anteriorly. Since we are dealing in sagittal plane, we can say that posterior movement would create extension and anterior movement would create flexion except at the knee. Also the LOG is creating the movement which is an external source; we can say that the joints experience external force due to gravity which creates flexion/extension moment depending upon the position of LOG. We have considered moment here which is a kinetic variable since it describes torque which is a force. In context to kinematics, the direction of motion should be considered.

Table 15.1 Kinematics of posture through position for LOG at different joint centers (cranial to caudal)

Joints	Position for LOG with respect to joint center	External moment due to LOG
Atlanto-Occipital	Anterior (passing through the external auditory meatus)	Flexion
Lower cervical	Posterior	Extension
Thoracic	Anterior	Flexion
Lumbar	Posterior	Extension
Lumbosacral/ Sacroiliac	Anterior	Flexion/Nutation
Hip	Posterior	Extension
Knee	Anterior	Extension
Ankle	Anterior	Dorsiflexion

15.3.2 Sagittal Plane Analysis of Sitting Posture

The sitting posture imposes higher contact stress and muscular activity due to shifting of LOG compared to the standing posture. Two distinct sitting postures include erect sitting and slumped or slouched sitting. Though slumped sitting seems comfortable, it is biomechanically incorrect and harmful. The LOG in slumped posture passes anterior to all segments of vertebral column compared to erect sitting. Thus the muscle activity of the posterior aspect would be higher to counter the external flexion moment. In erect sitting, the demand on the back extensor is also high to maintain the head over the trunk so that LOG passes closer to joint centers. The **flexion relaxation (FR) phenomenon** may provide a suitable base for suggesting that slumped sitting is comfortable than erect because after some time the muscle activity is very less due to flexion relaxation phenomena in back extensors. However the position is incorrect biomechanically.

15.4 Kinetics of Posture in Sagittal Plane

The most important force that acts on the human body determining the posture is Ground Reaction Force (GRF) acting on the center of pressure. At normal standing, the GRF lies along the LOG and it is responsible for creating the external moment. In response to the external moment, the muscles counteract by creating the internal moment in direction opposite to the external force created by the gravity. Table 15.2 represents the internal moment generated by the muscles against the external moment at different joint centers.

15.5 Pathomechanics

The deviation from the optimal posture can lead to various deformities affecting the kinematics and kinetics significantly. Let us understand the changes through malalignment in sagittal and frontal plane. The common deviations at joint centers

Table 15.2 External and internal joint moment in sagittal plane

Joints	External moment due to LOG	Internal moment
Atlanto-Occipital	Flexion	Extension by posterior neck muscle and ligaments
Lower cervical	Extension	Flexion by anterior neck muscle and ligaments
Thoracic	Flexion	Extension by ligaments
Lumbar	Extension	Flexion by abdominals and anterior ligaments
Lumbosacral/ Sacroiliac	Flexion/Nutation	Extension/Counternutation by ligaments and transverse abdominis
Hip	Extension	Flexion by iliopsoas
Knee	Extension	Flexion by hamstring and gastrocnemius
Ankle	Dorsiflexion	Plantarflexion by soleus

have been reported in Tables 15.3 and 15.4 for sagittal and frontal planes, respectively.

15.6 Nonpathological Postural Changes

The postural changes can also be seen due to nonpathological conditions such as aging, pregnancy, and sports. Let us understand in brief how these factors could affect the optimal posture.

15.6.1 Age Related Postural Changes

Aging is a continuous process that affects the integrity of musculoskeletal systems, thereby causing noticeable changes in the normal posture of human body. The muscle becomes weak and they generate insufficient force to overcome the gravitational moment which can shift the LOG from the optimal position. The bone also become weaker and their fracture can cause postural deformity as seen with the wedge fracture of the spine causing Kyphosis. The joint capsule and ligaments laxity can cause angular deviation such as femoral anteversion, tibial torsion, pelvic tilts, etc. In addition, the proprioceptive mechanism is affected in elderly which inhibits the sensory inputs and increases the tendency to lose balance and control among them. Therefore, older people are highly prone for increased risk of falls.

15.6.2 Pregnancy Related Postural Changes

Postural change in pregnancy is very commonly seen particularly as it advances. The weight of the growing infant shifts the COM and LOG significantly. The anterior portion of the abdomen is heavier which results to anterior pelvic tilt and thus the lumbar lordosis is increased.

Table 15.3 Postural changes in sagittal plane

Joints	Frontal plane changes	Biomechanical description	Kinematic and kinetic change
Ankle	Claw toes	Hyperextension of the metatarsophalangeal (MTP) joint, combined with flexion of the proximal interphalangeal (PIP) and distal interphalangeal (DIP) joints	Increased GRF at MTP leading to callus formation Reduction of BOS causing instability
Ankle	Hammer toes	Hyperextension of the MTP joint, flexion of the PIP joint, and hyperextension of the DIP	Altered length tension relationship of toe flexors and extensor
Ankle	Equinus	Fixed plantar position of ankle	Increased GRF at forefoot
Knee	Flexion contracture	LOG shifts posterior from anterior and creates an external flexion moment which should be resisted by Quadriceps LOG at hip goes anterior from posterior LOG for ankle shifts more anterior	Increased quadriceps activity leads to higher compressive stress Excess hamstring activity Excess soleus activity
	Genu recurvatum	Hyperextended knee posture LOG for ankle shifts more anterior	Excessive stress on posterior joint capsule and knee flexors Increased joint axial loading can cause degenerative changes
Pelvis	Anterior pelvic tilt	LOG shifts posterior	Increased Lordotic Curve at lumbar and cervical region, increased kyphotic curve at thoracic region
Lumbar spine	Increased lordotic curve	LOG shifts more posterior	Increased joint compression at facet joint and posterior aspects of the vertebral disk
Thoracic spine	Increased kyphotic curve	LOG shifts more anterior	Stretching of posterior ligaments and compression of vertebral body and anterior disk
Cervical spine	Forward head posture	LOG shifts more anterior which increased the normal lordotic curve	Increased compression on facet joint, posterior disk and narrowing of vertebral foramen Excessive activity of cervical extensors Affects temporomandibular joint function

Table 15.4 Postural changes in frontal plane

Joints	Frontal plane changes	Biomechanical description	Kinematic and kinetic change
Ankle	Pes planus or Flat Foot or Pronated foot	Medial longitudinal arch is collapsed and calcaneus is everted	Callus on medial side of the foot Peak plantar pressure increases
Ankle	Pes cavus or High Arch or Supinated Foot	Medial longitudinal arch is higher than normal and calcaneus is inverted	Excessive plantar pressure on the forefoot Plantar fascia gets excessively stretched Callus formation on metatarsal heads
Knee	Genu valgum or Knock Knee	Increased tibiofemoral angle	Stress on the medial longitudinal arch Abnormal weight bearing on the posterior-medial aspect of the calcaneus Close kinematic changes include flat foot, lateral tibial torsion, lateral patellar subluxation, and lumbar spine contralateral rotation
Knee	Genu varum or bow legs	Decreased tibiofemoral angle	Increased medial compressive force Medial translation of patella
Knee	Squinting or cross-eyed patella	Superior pole faces medially and the inferior pole faces laterally	Increased Q angle and lateral shifting of patella
Knee	Grasshopper-eyes patella	Superior pole faces laterally and the inferior pole faces medially	Abnormal patella tracking Instability of the patella
Hip	Pelvic hike	One side of the hip translates superior	Altered hip joint forces Hip abduction function is drastically affected
Vertebral Column	Scoliosis	Lateral flexion and rotation of vertebra	Asymmetrical growth and development of the vertebral bodies Compression of structure on the side of convexity
Shoulder	Shoulder drop	One side of the shoulder joint translates inferior with gravitational pull	Affects scapulohumeral rhythm Muscular imbalance Tensile force on the joint capsule and ligaments

15.6.3 Athletic Posture

The posture of an athlete is significantly affected by the nature and demand of the sports. For example, shooters attain a stiff shoulder which causes muscular imbalance in scapular muscles as per the demand of the training. On the other hand, a swimmer needs a mobile scapular muscle to perform his best. A study reported that shooters rely more on proprioceptive and vestibular component for postural stability to enhance their stance stability [4]. Similarly postural changes in trunk of equestrian could be assumed due to repetitive flexion and rotation activity. This can be supported by finding of a study that concluded significant functional asymmetry in lateral bending range of motion among horse riders [5].

15.7 Closed Chain Kinematic Analysis of Posture

It should be noted that changes in alignment of distal segments would lead to consequent changes in the distal segments as per the close kinematic chain relationship. Let us understand this relationship with the illustration as discussed below (Fig. 15.4). Consider that we analyze the kinematic-kinetic chain from distal segment (foot) to the most proximal segment (pelvis). It can be seen that pronation of foot (flat foot) can cause a compensatory internal rotation at Tibia which in turn causes a valgus knee. This could be further seen with compensatory internal femoral rotation with changes in the angle of anteversion. The femoral anteversion would cause excessive anterior pelvic tilt, which would cause increased lumbar lordosis. Similarly the biomechanical link changed can be seen up to head and neck with compensatory changes in the spinal column.

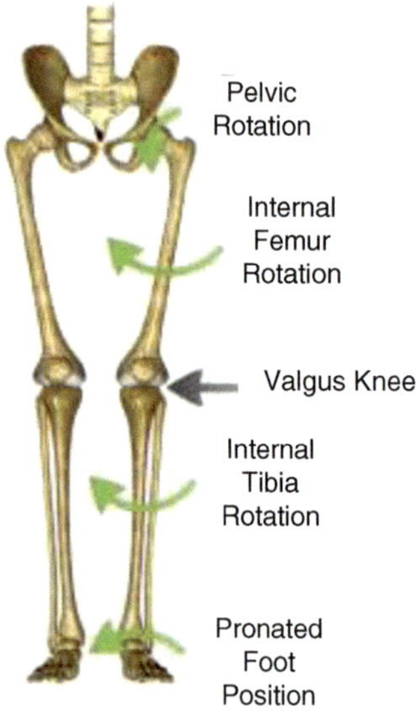

Fig. 15.4 Depicts close chain kinematic changes at knee and hip due to pronated foot

15.8 Summary

An ideal posture forms an important component of human biomechanics. An optimal standing and sitting posture is predominately determined by the LOG, COG, COM, and COP which allows efficient kinematics and kinetics at each joint segment. The alteration in the LOG would create significant changes in posture and lead to structural and functional changes.

References

1. Levangie PK, Norkin CC. Joint structure and function: a comprehensive analysis. Philadelphia, PA: FA Davis; 2011.
2. Maki BE, McIlroy WE. The role of limb movements in maintaining upright stance: the "change-in-support" strategy. Phys Ther. 1997;77(5):488–507.
3. Runge CF, Shupert CL, Horak FB, Zajac FE. Ankle and hip postural strategies defined by joint torques. Gait Posture. 1999;10(2):161–70.
4. Su FC, Wu WL, Lee WD. Stance stability in shooters. J Med Biol Eng. 2000;20(4):187–92.
5. Hobbs SJ, Baxter J, Broom L, Rossell LA, Sinclair J, Clayton HM. Posture, flexibility and grip strength in horse riders. J Hum Kinet. 2014;42(1):113–25.

Concepts of Kinesiology

16

16.1 Introduction

The term Kinesiology is well described as the study of human movement including body segments and organ system to promote better health. It could be defined as the study of the principles of mechanics and anatomy in relation to human movement [1]. It has been derived from a Greek word called kinesis meaning movement. It consists of physiological, psychological, and biomechanical domains. To restrict ourselves to the objective of this book we shall focus on the kinesiology aspects of biomechanics here. The human biomechanics which has been subdivided into kinematics and kinetics, their interrelation and interplay form the basis of kinesiology. We have learnt the simplified biomechanics in the previous chapters and focused on the linear kinematics and kinetics. In this section we will learn the concepts and principles of angular kinematics and kinetics which actually governs the kinesiology of human musculoskeletal system. The quantification for kinesiology is subject to mathematical operations. We shall learn the physical concepts and principles of kinesiology which is applicable to human biomechanics in common practice.

16.2 Relationship Between Linear and Angular Kinematics

The simplest form to understand the relationship between the linear and angular kinematics is to consider movement of body segments in angular motion and its applied motion on an object in linear motion. Till now, we considered the human joints as single segments and rigid body moving linearly; however, in reality the kinesiology of human biomechanics is combination of linear and angular. Let us understand this with an example below.

In a golf swing, the shoulder, elbow, and wrist move in angular kinematics and transfer it to the ball through linear (straight on the ground) or projectile motion (in air) as illustrated in Fig. 16.1.

A. Hazari et al., *Conceptual Biomechanics and Kinesiology*,
https://doi.org/10.1007/978-981-16-4991-2_16

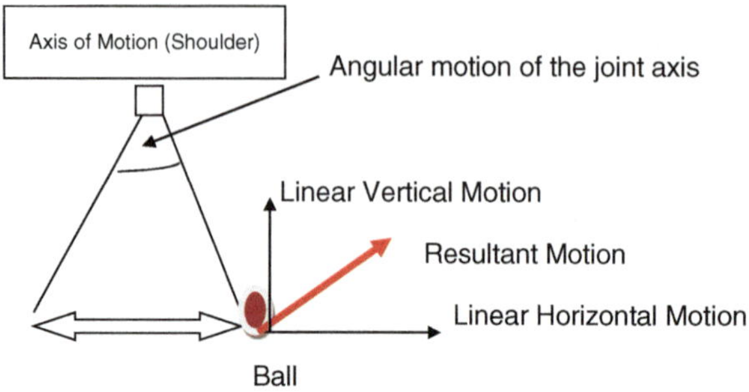

Fig. 16.1 Depicts application of angular kinematics and linear kinematics

So we see that the motion of human body segment is not pure linear, rather it is angular where the joint axis and their angulations have a major role to play. The angular kinematics is commonly represented by the range of angular motion, angular displacement, angular velocity, and angular acceleration.

The relation of angular and linear kinematics can be well understood when an object moves along the axis of body segment. Consider, you are holding a glass of water in your hand; you lift the glass and then place in your mouth to drink it. The wrist, elbow, and shoulder joints move in angular motion whereas the glass would show linear motion described as linear displacement.

16.2.1 Angular Velocity and Angular Acceleration

We have now understood that the human joint performs angular motions. Thus the velocity with which it moves would be known as the angular velocity. The velocity is vector quantity which has magnitude as well as directions and described through axis and planes for human joint movements. Similarly the rate change of angular velocity is known as angular acceleration. When we perform dynamic activities, the muscles undergo uniform acceleration and deceleration making the movement smooth.

16.3 Kinetics of Linear Motion

We have learnt about the types of linear force system and their resolution, Newton's laws of motion, force of gravity, friction, etc. Now the application of linear kinetics is well identified as the impulse–momentum relationship as discussed below:

16.3.1 Impulse–Momentum Relationship

The Newton's second law of motion ($F = m \times a$) forms the basis of biomechanical application. The linear momentum of a body is a measure of its mass and its linear velocity. Now the equation of second law can be represent as:

$$\text{Force} = \text{mass} \times \text{acceleration}$$

$$\text{Force} = \text{mass} \times \text{change in velocity} / \text{time}$$

$$\text{Force} \times \text{time} = \text{mass} \times \text{change in velocity}$$

The product of force and time is known as the **impulse** and the product of mass and velocity is known as **momentum** and their application in kinesiology determines the **impulse–momentum relationship** [1] as described in example below.

If we try to apply the human biomechanics governed by the equation above, the mass of the body remains constant; the maximum change in velocity depends upon the muscular strength and physiological range of motion, now the only factor that can enhance the application of force is time (impulse). If we increase the time of application, the force can be magnified. This concept is used in shot put throw, javelin throw, etc. The shot putter applies a force to the shot for a long period of time in order to produce more impulse and hence a greater change inmomentum. Similarly a soccer player applies force to kick the ball with a longer contact time which will cause an increase in the momentum and the ball would travel longer and faster.

16.3.2 Principle of Conservation of Momentum

According to this, the momentum created in a body is transferred to another as it is conserved. The concept can be applied and understood with example below.

A cricket fast bowlers run a distance and generate a momentum through muscular mass and change in body segment velocity starting from the lower limb, transferred to trunk , then to shoulder and ultimately to the ball. Thus the momentum generated in the body is transferred to another without much loss and this is known as the principle of the conservation of momentum. In another example, we wish to open a door that requires more than a usual force. We use our larger body mass with velocity (run) and generate a momentum to bang on the door. During the collision the momentum created by our body would be transferred to the door and it would likely open.

16.3.3 Moment of Inertia

The inertia of a body is resistance to linear motion or rest. The moment of inertia of the body is its resistance to state of angular rotation or rest [1]. The moment of inertia is directly proportional to its mass and distribution of mass about the axis of rotation. In simpler words, if the body rotates at an axis when the distribution of mass is away from the axis, the moment of inertia is more. Thus during pull up exercise, it is difficult to lift the body up from a completely extended elbow compared to flexed elbow.

16.4 Kinetic of Angular Motion

We now understand that the body segments move in angular motion more often than linear. Thus the force created by muscle would also show angular characteristics with application of angle or angle of pull. In other words, the force that muscles actually produce in human body should be considered in context to its mass, the perpendicular distance from the joint axis, and also the angle at which it works. Therefore the joint and muscle forces in human biomechanics are not only the product of mass and acceleration rather the angle of pull is also a major factor. When a force acts perpendicular to the joint axis, the magnitude is maximum (since angle of pull, Sin 90° = 1) known as the **"torque."** The application of torque in human biomechanics has high significance. While kicking a football we generate larger torque at hip keeping the lower limb straight.

16.4.1 Direction of Angular Kinetics

As per the law of inertia, the body tends to remain in a state of static or dynamic equilibrium. The equilibrium of a body can be achieved when the sum of all forces acting upon it equals zero. This is possible when forces being vectors are represented positive or negative depending upon the director of their motion. As a rule in biomechanics, all forces in **clockwise** motion would be **negative** and all forces in **anticlockwise** motion are considered **positive.** We shall follow this in all sections ahead.

16.4.2 Internal and External Torque

The application of torque or moment is very common in human biomechanics. We have learnt that the internal moment or torque is created by the muscles against the external torque created by the force of gravity. In kinesiology, the magnitude of the torque would determine the application and direction of force as explained below.

Consider that we are performing biceps curl with a weight in hand. Now the internal torque generated by the biceps brachii muscles is product of its force (muscle mass and change in velocity) and the perpendicular distance of its attachment from the joint axis. Similarly the weight would create an external torque equal to the product of its mass and perpendicular distance from the point of application. The biceps would generate torque to pull the weight upwards (anticlockwise) and weight would generate the torque downwards (clockwise). The net torque would be the difference in the magnitude of the external and internal torque. The direction of motion would be determined by the sign of net torque. If muscle torque is more represented as positive (anticlockwise), the weight would move towards the biceps or vice versa (Fig. 16.2).

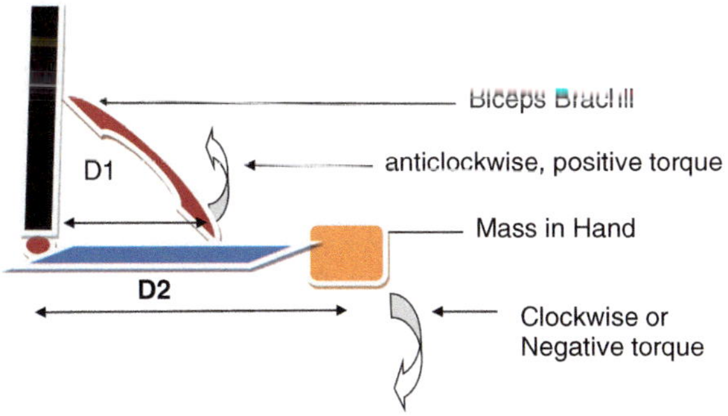

Fig. 16.2 Depicts internal and external torque while lifting a weight in hand (D1, moment arm of muscle; D2, moment arm of weight)

16.5 Center of Gravity and Center of Mass for Calculating Moment

The **center of gravity** of an object is defined as the point at which the entire weight of an object is assumed to be concentrated whereas the **center of mass** is the point at which the entire mass of the body is located. There are arbitrary points used for kinesiology [2]. For an evenly dense object the center of mass and center of gravity would lie at the geometrical center. If the mass and weight are not evenly distributed, the center of gravity and mass would shift to the heavier side. The knowledge of these points allows us to calculate the segmental kinematics and kinetics. The data on joint center location from the proximal and distal joints are given by David Winter [3]. This information is used for calculating the joint segment moment or torque as explained below.

Let us try to calculate the moment or torque produced by the muscle at elbow joint as shown in Fig. 16.2. Consider that the weight of human is 100 kg and distance for biceps attachment from joint center is .01 m. The mass of object in hand is 10 kg.

Now as we know,

$$\text{Moment} = \text{Force} \times \text{perpendicular distance}\left(d\right)$$

Let us first calculate the force using formula $f = m\,a$

Mass here would be that of forearm bone and mass of the object in hand. It has been proposed by Winter that the mass of the forearm is 2.8% of the body weight.

Thus $F = ((0.028 \times 100) + 10)$ kg $\times$ 9.81 m/s^2 (acceleration due to gravity in meter per second square)

$F = 100.9$ kg m/s^2

Now moment of torque at elbow = $F \times d$, i.e. = $100.9 \times 0.01 = 1.009$ Nm (Newton. meter)

16.6 Summary

The human biomechanics have been simplified to study as linear kinematics and kinetics. However in ideal situation the biomechanics follows angular kinematics and kinetics governed by the principles of kinesiology. Joint moment or torque is internal force governed by law of inertia, and momentum. The center of gravity and center of mass represent positions from joint centers for calculating the moment or force around an axis of rotation.

References

1. Grimshaw P, Fowler N, Lees A, Burden A. BIOS instant notes in sport and exercise biomechanics. New York: Garland; 2004.
2. Winter DA. Biomechanics and motor control of human movement. New York: Wiley; 2009.
3. Winter DA. Anthropometry. In: Biomechanics and motor control of human movement. Hoboken: Wiley; 1990.

Advanced Instrumentations and Their Interpretation

17

17.1 Introduction

The study of human movements has become very popular with the use of advanced technology and instrumentation in the last two decades. The most important components of human kinesiology include collection, interpretation of data and their use for sports performance and clinical rehabilitation. The advanced instrumentations have gained enormous popularity for biomechanical data collection and reporting with accuracy, which was not possible otherwise due to the nature of complex movements used by the human body. In this chapter, we shall learn about the latest instruments and their biomechanical output for interpretation used in sports and clinical biomechanics. The initial section would focus on the sports followed by clinical as well research. Multiple instruments are available. The instruments shown in this chapter are for reader reference as per the personal experience of the author and not for promotion.

17.2 Technology in Sports Biomechanics

Sports medicine has witnessed a technological revolution in diagnostic, clinical, and rehabilitation areas, which has taken the treatment in sports medicine to the next level. The most important areas where the technology in sports contributes are:

- Safer and more effective treatment
- Access to better preventive methods
- Stop the frequency of injuries or getting worse
- Faster Healing

A. Hazari et al., *Conceptual Biomechanics and Kinesiology*,
https://doi.org/10.1007/978-981-16-4991-2_17

17.2.1 Three Dimension Motion Analysis System

A powerful and useful tool in all sports for determining the fast movements including the technique. The most commonly used instruments are VICON, SIMI, QUALYSIS, BTS, etc. The 3D motion analysis systems have been discussed in detail under the clinical biomechanics section below. The important reasons for using the motion analysis system in sports include (a) Helpful in finding the cause of injury using Kinematics and Kinetics (b) In-depth insight about the body biomechanics, weight distribution and relative movement of joints and (c) analyze sports movement patterns for improvisation. Figure 17.1 shows motion analysis of a runner using high-speed infrared cameras and joint markers. The use of joint markers can be time taking and limited in their application. Thus motion analysis systems without the use of markers are available to perform dynamic sports analysis with accuracy (Fig. 17.2a, b).

17.2.2 Multi-Joint Dynamometer with Sports Simulator

Advanced instruments for isokinetic and isotonic muscle strength testing of various joints (Fig. 17.3)

17.2.3 Balance Assessment and Training System

A very useful instrument to train sports personnel where balance and coordination is required skill such as shooting sports (Fig. 17.4)

Fig. 17.1 Kinematic data recording for a runner using motion analysis system

Fig. 17.2 (**a**) Walking and Running analysis without joint markers (Walker View, Tecnobody, ABTP, Mohali, India, and Gulf Medical University Ajman UAE). (**b**) Running analysis without joint markers (D wall, Tecnobody, ABTP, Mohali, India, and Gulf Medical University Ajman UAE)

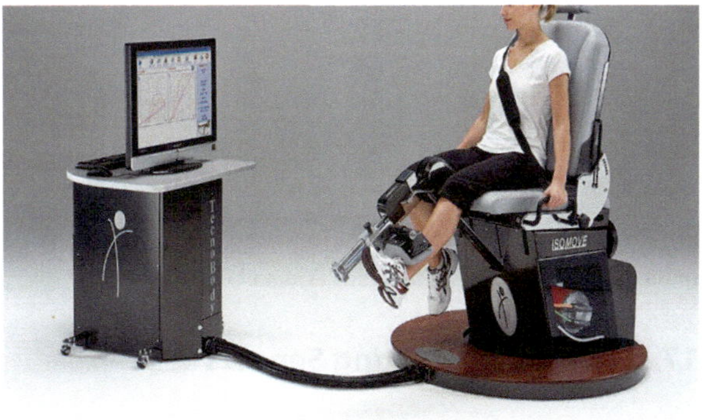

Fig. 17.3 Isokinetic dynamometer (Iso-Move, Tecnobody, ABTP, Mohali India and Gulf Medical University Ajman UAE)

17.2.4 Postural Control

The analysis of posture is a very important component in sports biomechanics. It allows us to identify the muscular imbalance and train accordingly (Fig. 17.5).

17.2.5 Cardiovascular Fitness Testing

The role of cardiovascular endurance training is of high importance in sports apart from muscular strength and skills. The cardiorespiratory strength and endurance are tested using advanced devices that perform Vo2 max analysis, metabolic testing and spirometry. The COSMED K4B2 and COSMED K5 are also widely used metabolic analyzers. Apart, PNOE is also the portable, affordable, and scientifically validated cardio-metabolic analyzer (Fig. 17.6).

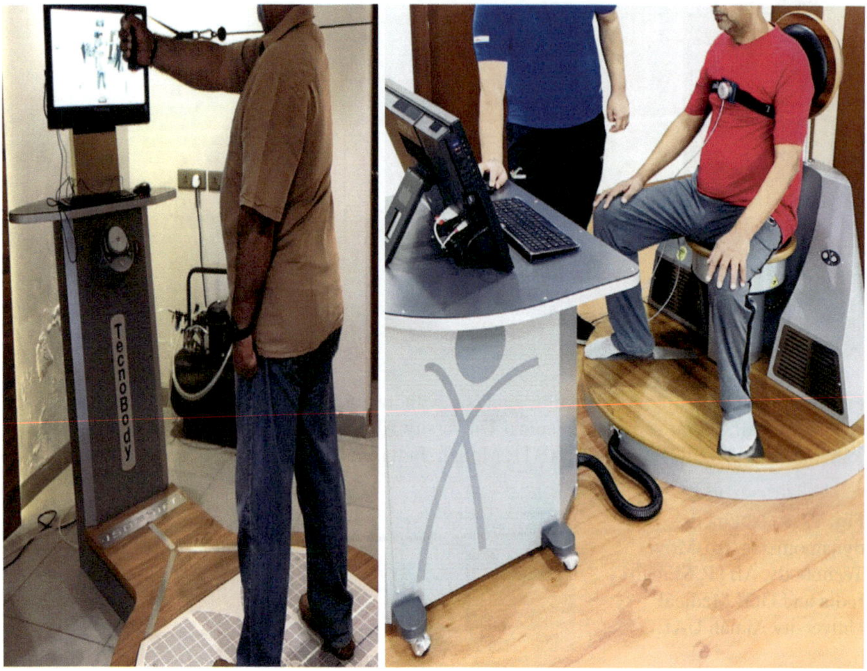

Fig. 17.4 Training for balance and coordination (Iso-Shift, Trunk MF, Tecnobody, ABTP, Mohali, India and Gulf Medical University Ajman UAE)

17.2.6 Injury-Detecting Sensors

The device is attached directly to the body or embedded in a wearable. For example, there are prototype sensors from **MotusPro** that can monitor arm movements in baseball and help players adopt the right movement ranges to avoid elbow injury. Similarly, head sensors in football detect how hard a player has been hit and calculate the risk for serious injury.

17.2.7 Early Brain Trauma Detection

The device allows detecting brain trauma on the field in real time. One of the devices known as **VEPS sensor system** comes in the form of a helmet-like device that can be used to detect changes in the patient's brain, which could signal brain injury. This enables quick treatment, which could prevent most of the negative effects associated with brain trauma.

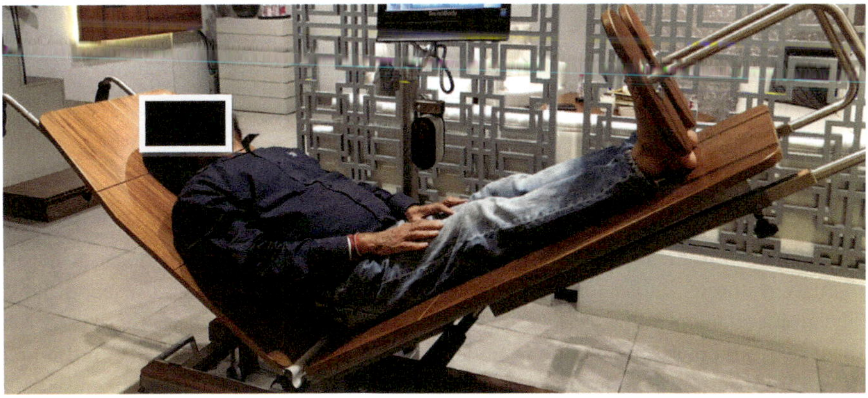

Fig. 17.5 Postural analysis (MF-PBENCH, Tecnobody, ABTP, Mohali, India)

Fig. 17.6 Metabolic testing for cardiovascular endurance (PNOE, ABTP, Mohali India)

17.2.8 Wearable

Wearable devices including smartwatches, smart clothing, wrist-worn bands, headgear, foot gear have been focused on health and fitness. These devices are loaded with sensors to monitor heart rate, keep track of daily activity, and monitor breathing and so on. One of the very useful devices for sports personnel is **Myontec** that produces intelligent clothing to monitors both the health and performance of an athlete. Others contain sensors to monitor things like pressure on joints, temperature, and movement range. It helps to warn the athlete when they are pushing their bodies too far and recommend the right movement ranges to prevent injury. One such device is **Smart Socks** equipped with pressure sensors in appropriate spots to help evaluate the possibility and extent of running injuries.

17.3 Technology in Clinical Biomechanics

The clinical biomechanics has been an area of continued advancement in instrumentation and technologies due to its demand in clinical setup and research labs. The biomechanical analysis in various pathological conditions is a matter of research for providing better knowledge in understanding the pathomechanics behind it and design appropriate management strategies. One of the areas that have been widely studied is Gait. Let us learn the technological advances in human gait analysis for clinical and research purposes.

17.3.1 Optoelectronic System for Motion Analysis

The motion analysis systems record data using multiple video frames and process it through digitization. More advanced optoelectronic motion analysis systems use infrared cameras and joints centers to yield 3 D kinematic and kinetic data. As mentioned above, the widely used systems are VICON, SIMI, QUALYSIS, BTS, etc. Retroreflective markers are placed on anatomical landmarks using various models. The very commonly used marker model for gait analysis is Helen HayesVicon *VCM* [1], *GaitLab* Peak [2], and Cleveland Clinic MAC *Orthotrak* [3]. The Fig. 17.7a, b, depict the QUALYSIS system set up for 3D motion analysis (Jupiter Hospital Thane, Maharashtra, India). The system consists of eight infrared cameras and two video cameras along with two force plates. This setup has been predominantly used to analyze pediatric gait such as cerebral palsy.

The SIMI motion analysis system is installed at the diabetic foot center, Kasturba Hospital, Manipal, Karnataka, India as shown in Fig. 17.8. The system is user friendly and has been extensively used for the kinematic analysis of the diabetic gait.

17.3.2 Static and Dynamic Pedography

Gait analysis is very commonly used for the clinical purpose in a variety of conditions. Since the optoelectronic system is very costly and requires skilled professionals to use it, the more user-friendly systems for clinical gait analysis include static and dynamic foot scanners, as shown in Fig. 17.9a, b. They provide useful spatio-temporal gait variables.

Fig. 17.7 (**a**) QUALYSIS motion analysis system set up at Jupiter Gait Lab, Jupiter Hospital Thane, Maharashtra, India. (**b**) QUALYSIS motion analysis system with embedded force plates at Jupiter Gait Lab, Jupiter Hospital Thane, and Maharashtra, India

17.3.3 Accelerometers

Acceleration of a body can be measured as the rate change of velocity. The accelerometers can directly measure the acceleration of the body when they are placed firmly. The most common accelerometers used in clinical and sports settings are either capacitive or piezoelectric. Triaxial accelerometers can provide the data in three planes and are very effective in clinical research.

17.3.4 Electromyography

The device is widely used to record muscle activity in both clinical and sports biomechanics. They detect the muscle signals using a pair of recording electrodes which are amplified to yield biomechanical variables like root mean square (RMS). **Myon AG** is one such device that can be used in portable mode (Fig. 17.10).

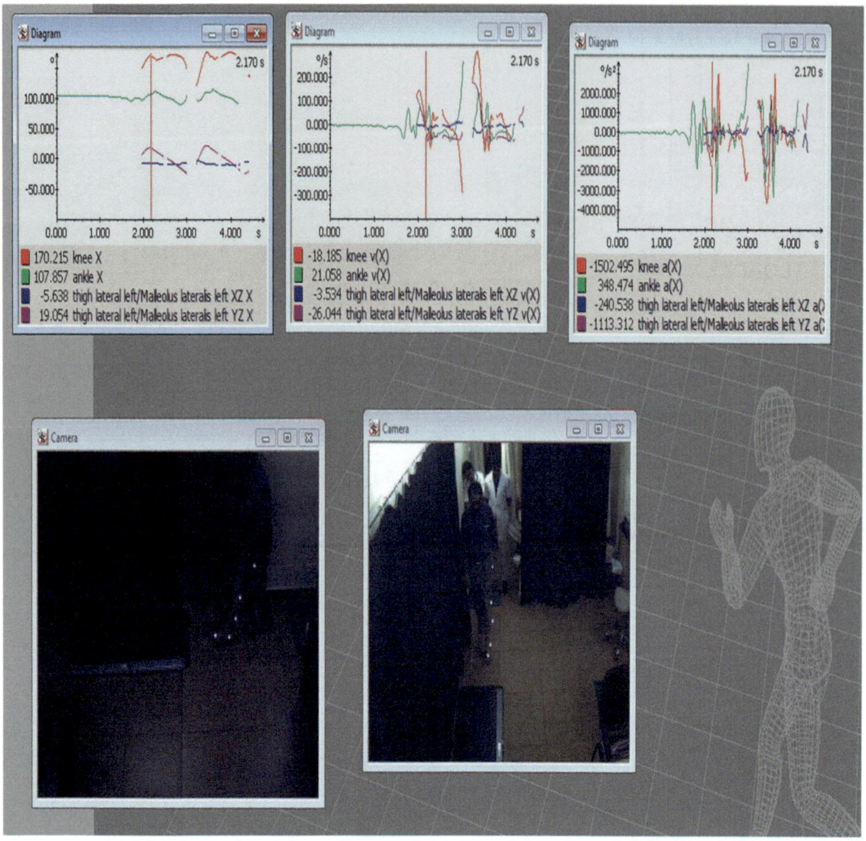

Fig. 17.8 Kinematic analysis of diabetic gait at Kasturba Hospital Manipal, Karnataka, India

Fig. 17.9 (**a**) I step static foot scanner (Aetrex, USA). (**b**) Wintrack dynamic Scan foot mat, Medicapteure software (France, USA) portable mode (Fig. 17.10)

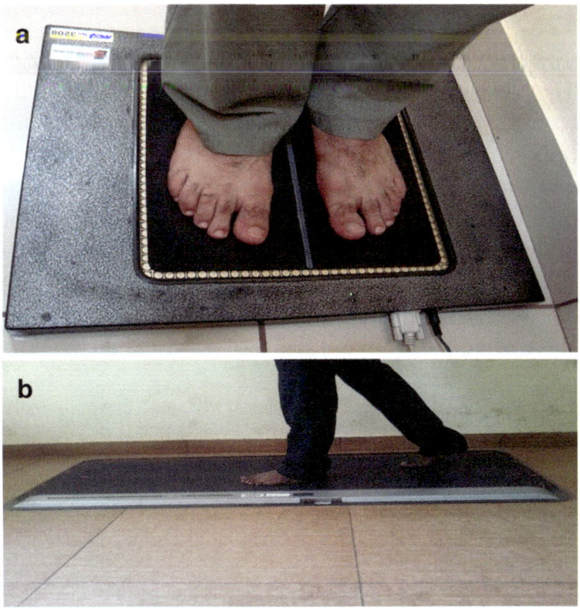

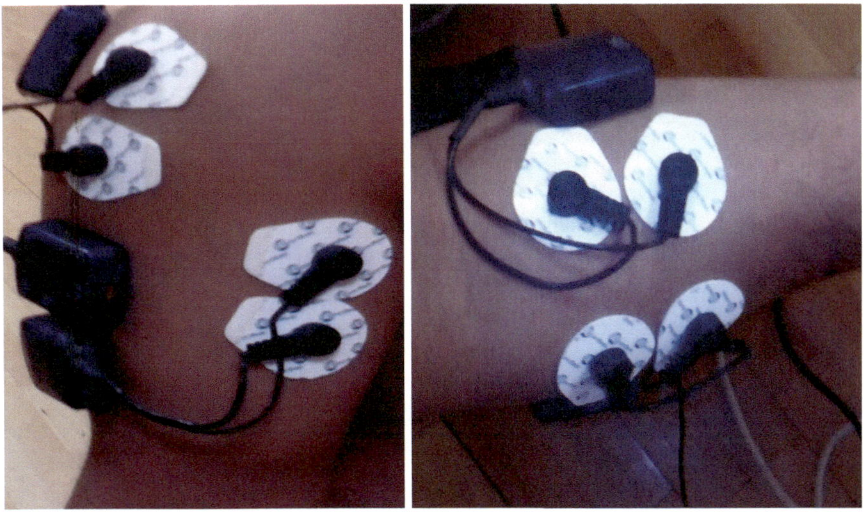

Fig. 17.10 Portable EMG electrodes placed on forearm and arm (myon AG, UK [4])

17.4 Summary

The advancement of technology has added value in the field of clinical and sports biomechanics. The advanced instruments can provide information in the shortest time with accuracy. The kinematic and kinetic data obtained can be used to determine pathomechanics and design suitable rehabilitation regimen. They can be highly useful in clinical and sports research. The use of such instruments should be promoted with cost-effective and user-friendly measures.

References

1. Davis RB III, Ounpuu S, Tyburski D, Gage JR. A gait analysis data collection and reduction technique. Hum Mov Sci. 1991;10(5):575–87.
2. Vaughan CL, Davis BL, O'Connor JC. Dynamics of human gait. Champaign, IL: Human Kinetics; 1992.
3. Kadaba MP, Ramakrishnan HK, Wootten ME, Gainey J, Gorton G, Cochran GVB. Repeatability of kinematic, kinetic, and electromyographic data in normal adult gait. J Orthop Res. 1989;7(6):849–60.
4. Hazari A, Warsi M, Agouris I. Electromyography analysis of shoulder and wrist muscles in semi-professional cricket fast bowlers during bouncer and yorker delivery. A cross-sectional comparative study. Int J Phys Educ Sports Health. 2016;3(6):77–87.

Applied Kinesiology in Sports 18

18.1 Introduction

The study of human movement in sports has been an area of interest in the current time. Since the sports have been very competitive and a marginal edge can determine the result of the competition, the biomechanics of sports have an important role to play. It has been widely used in analyzing the technique objectively; determine the outcome and quality of movement. The principles of biomechanics and their application in sports have drawn the attention of athletes as well as their coaches. Athletes are often keen to know the biomechanical strengths and skills required for sports, which can be easily determined by studying the characteristics of their movements. The players can also know their weaknesses and compare their performance with the competitors very objectively. Not only the athletes themselves, but the coaches also gain the advantage of improving their coaching skills through the applied kinesiology in sports. In this chapter, we shall learn kinesiology and biomechanical characteristics of popular sports like running, jumping, and throwing.

18.2 Biomechanics of Running

Running is a very popular sport involved in many athletic events. Thus it is very important to understand the biomechanics of running. Alike the normal walking gait, the running gait also consists of the stance phase and swing phase. However, the movement is very fast, and therefore a **period of flight** is seen when neither foot is in contact with the ground. The other noticeable change in running gait is the reversal of phase time contribution. In other words, the stance phase in the running contributes to 40% of the total gait cycle, and the swing phase covers the larger part of 60% [1]. The speed of running can be increased by either increasing the strike length or stride frequency. It has been proposed that the initial increase in speed

A. Hazari et al., *Conceptual Biomechanics and Kinesiology*,
https://doi.org/10.1007/978-981-16-4991-2_18

comes from increasing the stride length up to 7 m/s beyond which the stride frequency or cadence is used to increase the speed [1]. Increasing the stride frequency is, however associated with more energy consumption, thus makes the running less energy efficient.

18.2.1 Kinematics of Running

For the kinesiology of running, the stance and swing phase of running gait has been divided into sub-phases. The stance phase consists of **loading response, midstance, and drives off phase,** whereas the swing phase consists of **early swing and late swing.** The major kinematic events that take place during the running are highlighted below [1].

Hip Joint The range of motion for hip joint changes dynamically from 40 to 50° flexion at loading response to 10° hyperextension at drive off phase. In the swing phase, the hip joint moves from extension to 50° flexion at the late swing.

Knee Joint The knee is flexed to 60° during the loading response and extends by 20° at the drive off phase. During the swing phase and float period, the knee is maximally flexed to 125° at mid swing.

Ankle Joint The ankle joint reaches 10°–25° of dorsiflexion at the loading response and immediately pronates in the rest of the stance phase of running and as it enters the swing phase. The Plantarflexion reaches a maximum of 25° in the initial swing phase and then dorsiflexion of 10° in the late swing phase.

18.2.2 Kinetics of Running

Since running consists of a flight phase, the vertical movement of the body is greater, and thus the vertical velocity during the consequent foot strike of the stance phase is also higher [1]. The soft tissue would thus undergo higher stress and strain. As per the action-reaction phenomena given by Newton, the ground reaction force in the running would also be very high and expected to reach two–three times body weight [1]. Due to increased speed, the frictional force as well as the shear forces is also high. The knee joint in particular experiences high compressive joint reaction force due to vertical force components. The hip joint compressive force would be high due to the role of hip abductors to stabilize the hip during running and weight bearing on single leg as the period of double support is absent in the running. The rate of application of force is significantly affected in the running as the temporal parameters of gait are reduced due to faster moments and thereby affecting the viscoelastic properties of soft tissue significantly.

18.2.3 Energy Expenditure in Running

The expenditure of energy in running gait is very high compared to normal walking due to increased demand on the musculoskeletal system. Thus running biomechanics should include an energy efficient system. In ideal situations, as per the law of conservation of energy, the potential energy by virtue of its position is converted to kinetic energy by virtue of its motion. During running both energy are phased in such a way that when potential energy is high, the kinetic energy would also be higher. Thus the net energy, which is the sum of energy, makes the running energy efficient. The mechanism that has been identified which reduces the energy cost during running includes storage and return of energy in the elastic structures (Tendino-Achilles) and passive transfer of energy from one body segment to other [1].

18.3 Biomechanics of Jumping

Jumping is a popular sports event used by many athletes. Thus the kinesiology of jumping is highlighted here to understand its important biomechanical characteristics. Though there are many forms of jumping sports, we shall focus on the commonly applied biomechanics. The jumping sports can be biomechanically phased into **counter-movement, propulsion, flight, and landing** [1].

18.3.1 The Stretch-Shortening Cycle

It is described as the sequence of eccentric contraction applied through stretch followed by a quick concentric action to yield more power and force. The eccentric contraction of muscle produces a resistive force stored as the potential energy and referred to as the pre-stretch phase, which is quickly released as kinetic energy in the propulsive phase during the Counter-movement jump [2]. This is known as the stretch-shorten cycle. The faster the stretch and the shorter the delay, the greater will be the enhancement in the muscle force produced.

18.3.2 The Counter-Movement Phase

Most forms of jumping are initiated with a downward movement of the body referred to as a "counter-movement" which can help to achieve more jump height while utilizing the stretch-shorten cycle [1]. The knee, hip, and ankle flex in coordination to move the body into a better position to start the propulsive phase over a longer period of time. The basic principle is to generate a larger impulse during the propulsive phase as the force can be applied for a greater time. The more impulsive force would result in a higher takeoff velocity, and thus a greater jump distance can be achieved.

18.3.3 Jump Height

The height of vertical jump is directly proportional to the vertical velocity and measured as the square of takeoff velocity which is in turn related to the impulsive force.

18.3.4 Arm Swing in Jumping

Arm swing in jumping is an essential component that can add up to 20% of total jump height or distance [1, 2]. The swing of the arm is related to shifting the center of mass higher and generating a higher momentum.

18.3.5 Transfer of Momentum

It is the tendency of the body to rotate in accordance with the angular momentum during the flight phase of jumping [1]. The angular momentum is the product of the moment of inertia and velocity. For an effective transfer of momentum, the athletes purposely control the rate of rotation by elongation of the body and stretching out the limbs, which increases the moment of inertia and slows the rate rotation as per the law of conservation of momentum.

18.4 Biomechanics of Throwing

Apart from running and jumping, throwing events includes a variety of sports. It would be helping to understand the basic mechanics of throwing. The throwing sports involves common phase like **preparation, pulling, and follow through** [1].

18.4.1 The Preparation Phase

The major principle of the preparation phase is to generate a high impulsive force by determining the path of acceleration and pre muscle stretch. The preparation phase starts from the backward movement of the arm to the maximum horizontal position of the shoulder. The eccentric contraction in the anterior muscles of the shoulder facilitate the stretch-shortening cycle for generating high velocity.

18.4.2 The Pulling Phase

The major event that takes place during the pulling phase is to generate the velocity and attain the momentum in the lower segments of the body. The basic principle is to allow higher acceleration of the shoulder joint. During the early pull phase, the pelvis and trunk rotate backwards, which creates a rotational recoil force and stores the momentum through the body axis. The shoulder then rotates internally with elbow extension in the late pull phase. The momentum stored is transferred from the distal to proximal segments in a sequential manner and to the throwing object ultimately. The shoulder externally rotates to maximum [1].

18.4.3 Follow-Up Phase

The follow-up phase mainly involves deceleration of the body with a controlled stop. The shoulder muscles eccentrically contract to dampen the acceleration.

18.5 Summary

The kinesiology and biomechanical principles in sports can be applied in a variety of games. It is important to know the correct movement patterns of sports to yield better results. The study of sports biomechanics and kinesiology can be helpful in enhancing performance as well as rehabilitation during return to play.

References

1. Grimshaw P, Fowler N, Lees A, Burden A. BIOS instant notes in sport and exercise biomechanics. New York: Garland; 2004.
2. Graham-Smith P, Lees A. Science and football IV. Liverpool, UK: Research Institute for Sport and Exercise Sciences, Liverpool John Moores University; 2013. p. 29.

Biomechanics and Kinesiology in Strength and Conditioning

19

19.1 Introduction

Strength and conditioning have been an area of biomechanical interest in the current era where multiple researches are being conducted. The National Strength and Conditioning Association (NSCA) is world's reputed organization that provides a variety of academic and research information to sports personnel as well as their coaches. It designs various courses and certifications for training the professionals involved in strength and conditioning. The availability of evidence-based practice provides a scientific base for the professionals to prepare and plan their competition along with rehabilitation plans. It is the level of condition for an athlete to be categorized into premature, advanced, or elite. Since exercises are now being prescribed as medicine, it is important to know the kinesiology in strength and conditioning where we apply the biomechanical principles to achieve the desired outcome. For example, we understand that a heavy weight lifter's training principles vary from running as per his demands, and accordingly we train the athlete keeping the biomechanical structure and function in mind. In this chapter, we shall deal with biomechanical techniques of squat analysis, which is a good example of strength training. In addition, the plyometric conditioning for jump would be dealt. The work of Knudson in biomechanical analysis of various sports holds high significance in the kinesiology of strength and conditioning [1–3].

19.2 Technique Analysis for Squatting in Heavy Weight Lifters

Squatting is a functional movement that is used in a variety of sports. When we are dealing with heavy weight lifters, one of the most important components of technical analysis is mastering the squat. The squatting movement with free weight makes it a very important area of biomechanical analysis because small dysfunction or

A. Hazari et al., *Conceptual Biomechanics and Kinesiology*,
https://doi.org/10.1007/978-981-16-4991-2_19

mistake can cause heavy damage to the musculoskeletal system. Apart, the proper technique can help to perform that movement efficiently without risk of injury. The kinematic and kinetic analysis of squats can help determine and enhance the performance-related parameters of weight lifting. The important biomechanical aspects of a good weight lifting technique with squats have been highlighted as below [4].

(a) Keeping the stance width appropriate, ideally equal and below the shoulder width.
(b) Balance of the bar over the body and maintaining the COM within the base of support.
(c) Smooth and coordinated close kinematic movements of the spine and lower limb joints.
(d) Performing the squat slowly and smoothly using the Force-Time Principle where the stress on muscle is efficient.
(e) Keeping the spine straight such that the compressive loads on the disk are evenly distributed. Studies have shown that flexed spine interferes with the resistive capacity of the spine extensor for controlling the anterior shear force and thus making the spine unstable [5].
(f) Keeping the LOG closer to the joint axis would stress the muscles less against the external moment. The technique should also be performed within the appropriate ranges of motion. Hyperflexion at the knee should be avoided to control exaggerated shifts of LOG.
(g) The optimal trunk leans with hip flexion have been identified as the primary factor in determining the distribution of joint moments that contribute to the exercise [6–8]. An overtly straight trunk would reduce the length-tension relationship of extensor muscles and thus reduce the extensor torque at the hip and lumbar segments, whereas the demand on knee extensor would increase significantly.

It is important to note that the sports coaches are trained with the understanding of biomechanical principles of technical analysis which they should incorporate in their coaching sessions. The coaching for any sports has not become a quantifiable outcome that should be a part of the academic and research curriculum.

19.3 Plyometric Training

Plyometric is a form of muscle strength and condition where the principle of **stretch-shorten cycle** is used to generate power and more force. Recent studies prove that jumping abilities are enhanced when the body is conditioned well to use plyometrics [9]. The technique of jump analysis is important to perform the movement efficiently without risk of major injuries as the involvement of impact forces from the ground makes it highly vulnerable to musculoskeletal injuries. Since resistance training has been shown to provide positive effects in jumping techniques

[10], it would be highly useful for coaches and trainers to understand the principle of resistance training in jumping sports.

The important biomechanical aspects of a good jumping technique have been highlighted as below.

(a) The position of the lower limb joints to provide a good stretch-shortening cycle and increase the jump height—The role of stretching in sports is well studied and holds a significant contribution in sports performance [4, 11].
(b) A good countermovement jump with significant vertical takeoff velocity.
(c) Utilization of arm swing to enhance the transfer of momentum and reduce the loading force on lower extremity.
(d) Strengthening of lower extremity muscle for powerful force generation. The muscles like the quadriceps should be used as a powerful knee extensor for a vertical lift. Similarly, the Soleus should be an efficient shock absorber for ground contact.
(e) Resistive and plyometric training using external weight, medicine ball, drop jump etc., can improve the performance.

19.4 Principle of Specificity in Strength and Condition

It is evident to understand that the demand of the sports determines the type of input for training. This is best understood with the **principles of specificity.** The principles state that when strength training is required, then the load should be used as the variable for conditioning and when the endurance is needed, then volume should be altered. In simple words, all sports that deal with strength as their main focus, such as weight lifting, we should use external weights for training, whereas when endurance needs to be built (marathon running), the frequency or the number of sets should be increased gradually to attain a higher conditioning level.

19.5 Summary

Strength and conditioning is a state of expertise where biomechanical principles play an important role. Understanding the correct biomechanics and technique analysis can provide a better outcome for sports performance, training, and coaching.

References

1. Knudson D. The magnus effect in baseball pitching. In: Wilkerson J, Ludwig K, Butcher M, editors. Proceedings of the 4th national symposium on teaching biomechanics; 1997. p. 121–5.
2. Knudson D. Biomechanics of the basketball jump shot: six key teaching points. JOPERD. 1993;64(2):67–73.
3. Knudson D. The tennis topspin forehand drive: technique changes and critical elements. Strategies. 1991;5(1):19–22.

4. Knudson D. Fundamentals of biomechanics. New York: Springer; 2007.
5. McGill SM, Hughson RL, Parks K. Changes in lumbar lordosis modify the role of the extensor muscles. Clin Biomech. 2000;15:777–80.
6. Escamilla RF. Knee biomechanics of the dynamic squat exercise. Med Sci Sports Exerc. 2001;33:127–41.
7. Hay JG, Andrews JG, Vaughan CL, Ueya K. Load, speed and equipment effects in strength-training exercises. In: Matsui H, Kobayashi K, editors. Biomechanics III-B. Champaign, IL: Human Kinetics; 1983. p. 939–50.
8. McLaughlin TM, Lardner TJ, Dillman CJ. Kinetics of the parallel squat. Res Q. 1978;49:175–89.
9. Bobbert MF. Drop jumping as a training method for jumping ability. Sports Med. 1990;9:7–22.
10. Hunter JP, Marshall RN. Effects of power and flexibility training on vertical jump technique. Med Sci Sports Exerc. 2002;34:478–86.
11. Knudson D. Stretching: science to practice. JOPERD. 1998;69(3):38–42.

Calculating Muscular and Joint Forces 20

20.1 Introduction

With the advanced technologies, the need for complex mathematical calculations for biomechanical analysis is very efficiently met. However, manual calculation of the kinetic parameters like muscular and joint forces acting on the body provides insight for an appropriate understanding of the biomechanical laws and principles. Analysis of these forces would not only quantify the forces acting on the body, but it will also give us in-depth knowledge of the linear and concurrent force systems. The human body has two major force vectors. One is rotatory, and the other is translatory. When multiple forces act on the joint segment, the net force needs to be determined using the biomechanical concepts and Newton's laws of motion. In this chapter, we shall learn how to calculate the muscle and joint forces on human body segments, and the same concepts can be applied to different segments.

20.2 Trigonometric Applications in Human Biomechanics

The use of trigonometry forms the fundamental basis for the calculation of forces in human biomechanics. We know that the forces are vector quantities and thus would have magnitude and direction both. In addition, the forces acting on the body are applied at a different angle due to the orientation and attachment of the muscles to the joint. Thus the use of trigonometry can help us to calculate them, as explained below.

20.2.1 Vectors and Triangle in Human Body

For simplification, the forces acting on the human body can be considered as a right angle triangle where one side is formed by the rotatory force component, the other

A. Hazari et al., *Conceptual Biomechanics and Kinesiology*,
https://doi.org/10.1007/978-981-16-4991-2_20

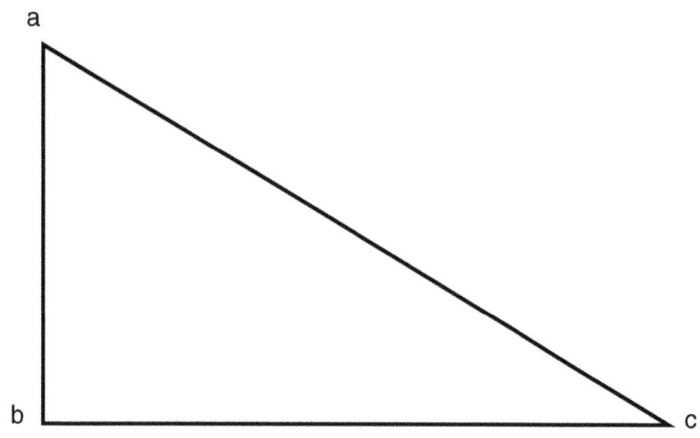

Fig. 20.1 Sides of the triangle (ab—perpendicular, bc—base, and ac—hypotenuse)

side represents the translatory force component, and the third side is represented by the resultant force vector. The resolution of the force can be done using the law of triangle or parallelogram law, as explained in the initial chapters. The sides of the triangle are explained in Fig. 20.1. The side along the x-axis is known as the base of the triangle; the side along the y-axis is called the perpendicular and the side joining the x-axis and y-axis is the hypotenuse.

20.2.2 Pythagorean Theorem

According to the theorem, when two forces act perpendicular to each other as part of the right angle triangle, the resultant is expressed as the diagonal of the triangle, which can be measured as the square root of the sum of perpendicular forces. In simple words, $ac^2 = ab^2 + bc^2$. Thus the law of Pythagoras can be applied to human biomechanics to find the resultant force when the magnitude of the rotatory and translatory force is known.

20.2.3 Laws of Trigonometry Angle in Human Kinetics

The application of Vector analysis through the Pythagoras theorem is very simplified for human kinetics since it only considers the forces acting perpendicular to each other or at 90°. In the real situation, the force of muscle acting on the human joints is not part of right angle triangle always. Rather the angle of pull is varying. Thus to calculate the exact magnitude of forces acting on the body, the angle of the force is a very crucial component and should be measured using the laws of trigonometry angles (Sine, Cosine, Tan).

Consider the hip joint in Sagittal view as shown in Fig. 20.2. If we closely look at the orientation of the femur, it is not exact vertically down, rather it is attached to the hip and knee at an angle anterior to the hip joint axis. Let us assume that the attachment of the femoral head to the hip joint makes an angle of 30° with the

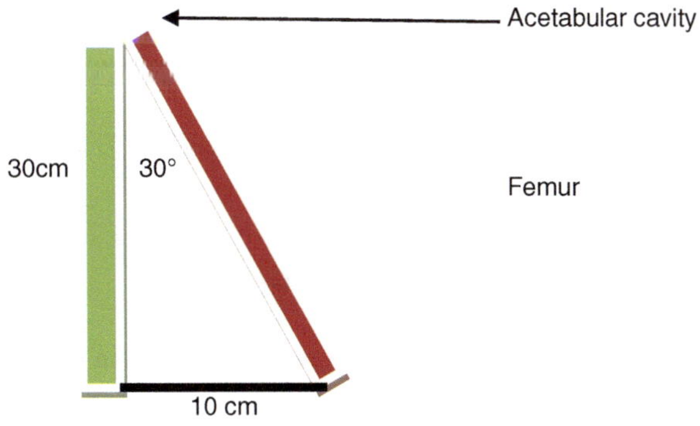

Fig. 20.2 Illustration of hip articulation with femoral angle and moment arm

vertical axis, as shown in Fig. 20.2. The distance of the distal end of the femur from the vertical axis (represented by green line) is 10 cm, and the vertical distance from the hip joint is 30 cm.

Considering the triangle above,	Perpendicular side (opposite side) = 10 cm
	Base side (adjacent side) = 30 cm
	Angle = 30°

What would be the size of the hypotenuse that would represent the length of the femur?

Using Law of $\text{Sine}\,\theta = \dfrac{\text{Perpendicular}}{\text{Hypotenuse}}$

$$\text{So Sine}30^\circ = \frac{10}{\text{Hypotenuse}}$$

$$\text{Therefore Hypotenuse} = \frac{10}{\text{Sine}30^\circ}$$

Since value of Sin 30° = 0.5

Therefore, Hypotenuse = 10/0.5 = 20 cm

The easy way to remember the laws of trigonometric angles is to memorize the following phrase:

Some **P**eople **H**ave **C**urly **B**rown **H**air **T**ightly **P**ulled **B**ack

Where,

- S = Sin Đ
- C = Cos Đ
- T = Tan Đ
- P = Perpendicular side
- B = Base side
- H = Hypotenuse

Such that

- Sin Đ = Perpendicular / Hypotenuse
- Cos Đ = Base / Hypotenuse
- Tan Đ = Perpendicular / Base

20.3 Resolution of Vectors

The resolution of vectors has been dealt with in the initial chapters, where the vectors follow the principles of addition in the same direction and subtraction in the opposite direction. However, when vectors act at angle , the use of parallelogram law and trigonometric function are done to resolve the vectors, as shown in Fig. 20.3. We know that ground creates and equal and opposite force to our body weight which is known as the ground reaction force vector (GRFV). The GFRV passing at a certain distance from each joint segment creates a significant angle (Fig. 20.3a). Now we know that the resultant force acting of a joint segment can be resolved into the horizontal rotatory and vertical translatory component at the human joints (Fig. 20.3b). Thus if the resultant GRFV has a magnitude of 500 Newton acting at 80° from the horizontal plane, we can resolve the forces using the trigonometry application and completing the triangle such that the resultant force acts as the hypotenuse (Fig. 20.3c). Now, if we can find the magnitude of the base and perpendicular, we can determine the horizontal and vertical force components, respectively.

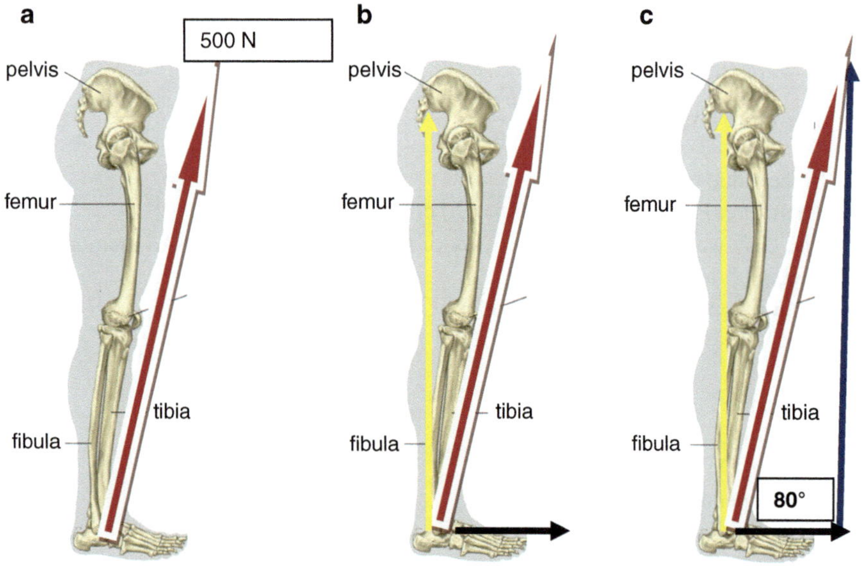

Fig. 20.3 (**a**) Showing ground force vector (**b**) showing horizontal and vertical force vector (**c**) resolution of force vectors uses the law of triangle and trigonometry

Given that,	the resultant force = 500 N
	Angle = 80°

Now, we can determine the vertical force component using Sine law as below.

$$\sin 80^\circ = \frac{\text{Perpendicular}}{\text{Hypotenuse}}$$

Therefore,
Sin 80° × Hypotenuse = Perpendicular
0.98 × 500 = Perpendicular or
Perpendicular = 490 N

Similarly, we can find the horizontal force component using the law of Cosine angle as below.

$$\cos 80^\circ = \frac{\text{Base}}{\text{Hypotenuse}}$$

Therefore,
Cos 80° × Hypotenuse = Base
0.17 × 500 = Base or
Base = 85 N

Implication According to the hypothetical example above, a resultant ground reaction force of 500 N produces a propelling or forward force of 85N and an upward reaction force of 490 N against body weight.

Note We have resolved the horizontal and vertical force components from the resultant vector here. In cases where these two force vectors are known, and we need to know the resultant force, we still can apply the same law of triangle/parallelogram and Pythagoras theorem along with trigonometry as applicable.

20.4 Resolving Muscular Forces

Having learned the trigonometric application and resolution of vectors, we can now learn how to resolve the muscular forces on the joint. This is a simplified calculation for your understanding. Let us analyze and measure the force of hip abductors and adductors on the femur using the vector diagram 20.4 below. Consider that the hip adductor creates a net resultant force of 1000 N, whereas the hip abductor creates 800 N forces when attached at an angle of 40° and 10°, respectively (Fig. 20.4). There are two methods to use the reference frame for resolving force vectors **(a) Resolving the component of force at 90° to the body segment, or (b) Resolving the horizontal and vertical components of the force relative to the ground** [1, 2]. In this book, we shall use the first reference. Therefore it should be noted that the muscle force is resolved in respect to the long axis and perpendicular axis of the

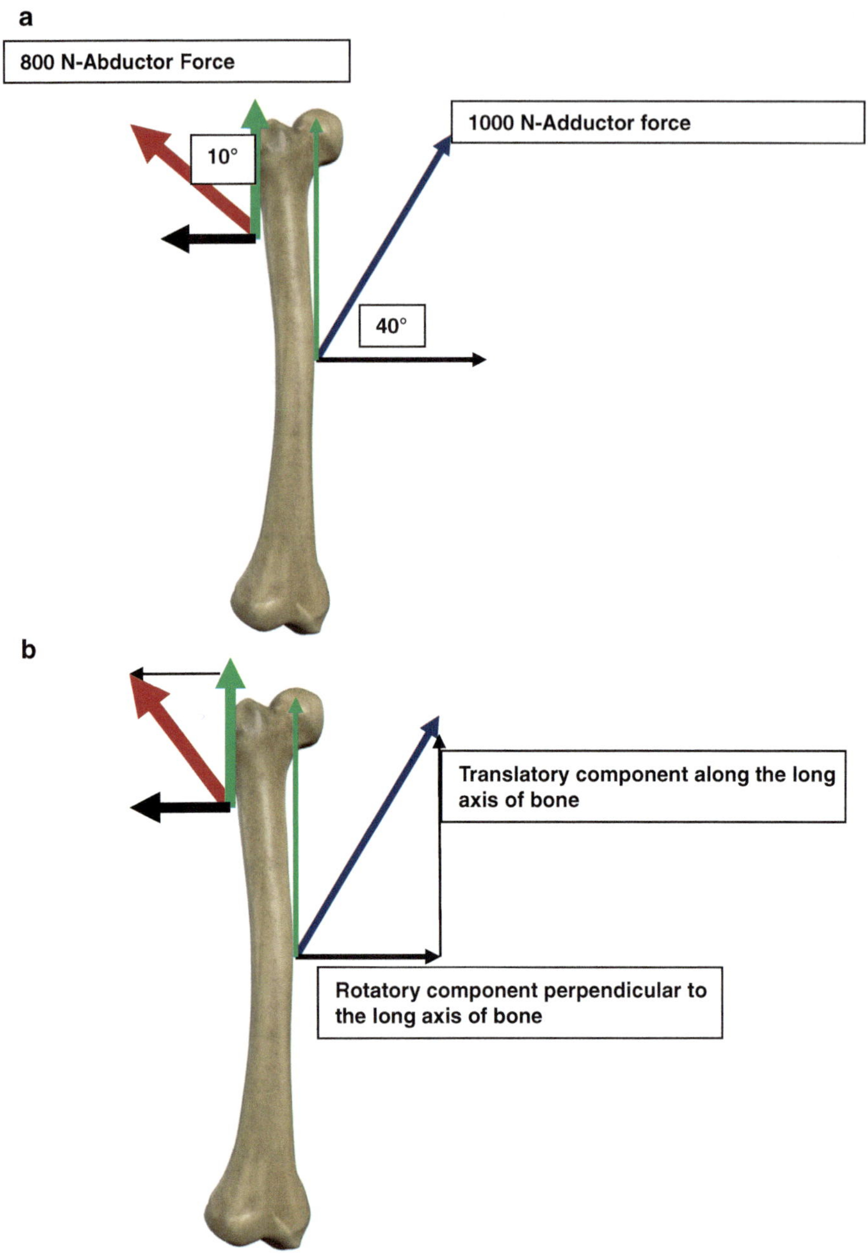

Fig. 20.4 (**a**) Depicts net hip adductor and abductor force vectors along the long and perpendicular axis of the bone. (**b**) Resolution of hip adductor and abductor force vectors along the long and perpendicular axis of the bone

bone, as shown by vector diagrams (Fig. 20.4b). For calculating the force of the hip adductor and abductor, let us complete the triangle as shown in Fig. 20.4b, and use the laws of trigonometry.

Calculating the Hip Abductor Force Components
Rotatory Force:

$$\sin 10^{\circ} = \frac{\text{Perpendicular}}{800\text{N}}$$

Therefore Perpendicular or Rotatory force = Sin 10° × 800 = 136 N
Translatory Force:

$$\cos 10^{\circ} = \frac{\text{Base}}{800\text{N}}$$

Therefore Base or Translatory force = Cos 10° × 800 = 784 N

Calculating the Hip Adductor Force Components
Translatory Force:

$$\sin 40^{\circ} = \frac{\text{Perpendicular}}{1000\text{N}}$$

Therefore Perpendicular or Translatory force = Sin 40° × 1000 = 640 N
Rotatory Force:

$$\cos 40^{\circ} = \frac{\text{Base}}{1000\text{N}}$$

Therefore Base or Rotatory force = Cos 40° × 1000 = 766 N

Note The magnitude of the force is considered positive and negative based on the Cartesian coordinate system. Therefore all forces acting upwards are positive, and downwards forces are negative. All forces acting towards the right are positive, and left are negative.

$$\begin{aligned}\text{Total muscle force at hip joint} &= \text{hip abductor force} + \text{hip adductor force}\\ &= -136 + 784 + 640 + 766 = 2054\,\text{N}\end{aligned}$$

20.5 Calculating Moments and Force Around a Joint

In the above sections, we have learned the use of trigonometry and resolution of force vectors through a simplified body analysis. However, in real situations, the calculation of forces at the joint is quite complex. Since the attachment of muscles on the joints acting as the pivot creates rotational moments, it is important to understand the concepts of moments before calculating the forces on the joint. We have mentioned multiple times that in order to be stable or in a state of static and dynamic equilibrium, the external moment should be equal and opposite of the internal moment such that the sum of the force acting at the joint is equal to zero. We shall use this concept in our calculation henceforth. The moment of a body segment is determined by its mass and center of mass around which it rotates predominantly. The magnitude of moment for a body segment would be its product of mass (force) and perpendicular distance from the center of mass (moment arm or radius of gyration). The calculation of the net force on the joint has three major components [1, 2].

1. Balancing force of muscle (Internal muscle moment)
2. Joint force from the muscle and the segment weight (Horizontal and Vertical Component)
3. Joint moments due to gravity (External moment)

20.5.1 Calculating Anthropometric Data

Since the anthropometry between human differs a lot based on the race and demographics, a cadaveric experiment performed by Dempster [3] formed the basis of standard anthropometric data for the calculation of dynamic moments as listed in Table 20.1.

Table 20.1 Anthropometrics of body segments

Body segments	Length of the segment with respect to body height	Mass of segment with respect to body mass	Center of mass (CoM) with respect to segment length (measured from proximal end)	Radius of gyration with respect to segment length (measured about CoM)
Foot	0.152 (length) 0.055 (width)	0.014	0.429	0.475
Shank	0.246	0.045	0.433	0.302
Shank and foot	0.285	0.0595	0.434	0.416
Thigh	0.245	0.096	0.433	0.323
Entire lowerextremity	0.72	0.157	0.434	0.326
Upper arm	0.186	0.0265	0.436	0.322
Forearm	0.146	0.0155	0.430	0.303
Hand	0.108	0.006	0.506	0.297
Forearm and hand	0.254	0.0215	0.677	0.468
Entire upper extremity	0.441	0.0485	0.512	0.368
Head and trunk minus limbs	0.52	0.565	0.604 (from top of head)	0.503

The information given in Table 20.1, can be used to find the essential information like segment length, segment mass, the center of mass and radius of gyration, which can further be used to calculate the segment moment and joint force based on individual's height and body mass.

Let us try to find the anthropometry data of thigh segment using the information given in the table for an individual weighing 70 kg and height of 1.5 m.

$$\text{Length of the segment}(\text{thigh}) = 0.245 \times 1.5(\text{body height}) = 0.367\,\text{m}$$

$$\text{Mass of the segment}(\text{thigh}) = 0.096 \times 70(\text{body weight}) = 6.72\,\text{kg}$$

$$\text{Center of mass}(\text{thigh}) = 0.433 \times 0.367(\text{segment length}) = 0.158\,\text{m}$$

$$\text{Radius of gyration}(\text{thigh}) = 0.323 \times 0.237(\text{segment length}) = 0.076\,\text{m}$$

Note: the calculation for anthropometry has been shown because the force on the human body is measured at different joint segments as free body analysis and not as a single body.

20.5.2 Calculating the Net Joint Force

We have learned the resolution of muscle force in the Sect. 20.4 above. For calculating the joint forces, all forces acting in the vertical direction, including muscles and weight of the body segment, would constitute the vertical or translatory force component, and all forces acting in the horizontal direction will constitute the horizontal or rotatory component. The positive and negative force would be determined as per the coordinates system discussed above. Since the horizontal and vertical joint forces are perpendicular to each other, acting at 90°, we can use the Pythagoras theorem to find the resultant joint force using the formula below:

$$\text{Resultant joint force}^2 = \text{vertical joint force}^2 + \text{horizontal joint force}^2$$

20.5.3 Calculating the Joint Moments

The joint moment is a rotating force, also known as the torque produced by the magnitude of ground reaction force and the perpendicular distance of the line of gravity from the joint center. The direction of the joint moment in clockwise is considered negative, and anticlockwise is considered positive. Let us consider the example shown in Fig. 20.5a below to determine the joint moments at ankle, knee, and hip. We can see that the GRFV pass posterior to the ankle, posterior to the knee and anterior to the hip such that it would create an external plantarflexion at the ankle, flexion at the knee and extension moment at the hip (clockwise movement of the femur thus negative). The moment arm for each joint is given in Fig. 20.5b. Let us consider the following situation for the calculation of external moments at each joint.

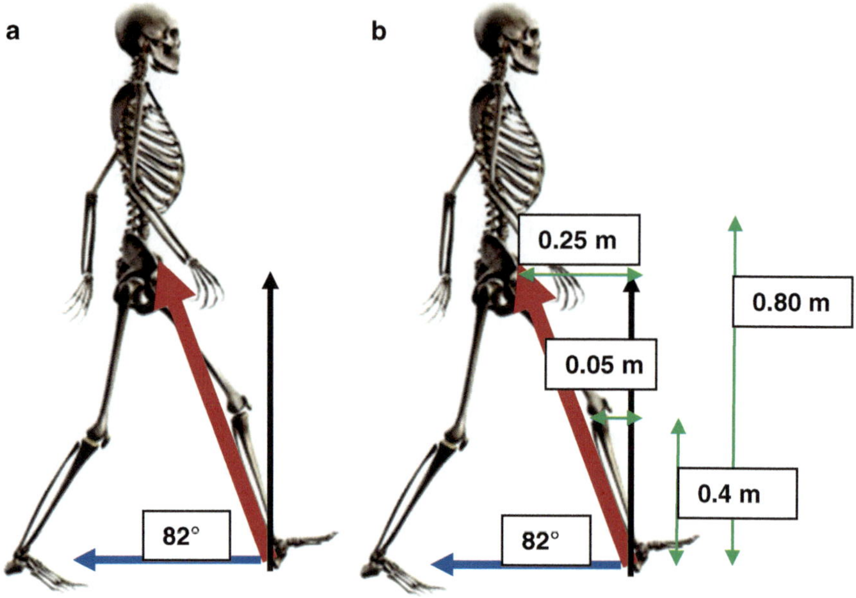

Fig. 20.5 (**a**) Depict ground reaction force acting at ankle, knee, and hip. (**b**) Depict example for horizontal and vertical moment arm at knee and hip

- The magnitude of GRFV = 950 N
- Angle of GRFV application- 82°
- Perpendicular Horizontal distance from the ankle joint = 0.1 m
- Perpendicular Vertical distance from the ankle joint = 0 m as GRFV passes through the joint axis.
- Perpendicular Horizontal distance from the knee joint = 0.40 m
- Perpendicular Vertical distance from the knee joint = 0.08 m
- Perpendicular Horizontal distance from the hip joint = 0.85 m
- Perpendicular Vertical distance from the hip joint = 0.25 m

Resolving GRFV:

$$\text{Horizontal ground reaction force} = 950 \times \cos 82^{\circ} = 132.2\,\text{N}$$

$$\text{Vertical ground reaction force} = 950 \times \sin 82^{\circ} = 940.75\,\text{N}$$

Now, resolving moments at each joint

Ankle

$$\text{Horizontal ground reaction force} = 132.2\,\text{N} \times 0.1\,\text{m} = 13.2\,\text{Nm}$$

$$\text{Vertical ground reaction force} = 940.75\,\text{N} \times 0 = 0$$

Knee

$$\text{Horizontal ground reaction force} = 132.2\,\text{N} \times 0.40\,\text{m}$$
$$\text{Vertical ground reaction force} = 940.75\,\text{N} \times 0.08\,\text{m}$$

Hip

$$\text{Horizontal ground reaction force} = 132.2\,\text{N} \times 0.85\,\text{m}$$
$$\text{Vertical ground reaction force} = 940.75\,\text{N} \times 0.25\,\text{m}$$

We find that the GRFV creates a significant external joint moment at each joint which should be opposed by the internal moment created by muscular force. At the ankle joint we found that the GRFV created a plantarflexion moment which should be counteracted by the dorsiflexion moment by the tibialis anterior muscle. If we wish to find the muscular force to be generated by the tibialis anterior to counter the external moment, we can use Newton's law of equilibrium to calculate it as below.

If the distance of tibialis anterior from the ankle is 0.01 m, then as per Newton's law

$$\text{Tibialis anterior force} \times 0.1\,\text{m} + \text{external moment at ankle}$$
$$= 0\,(\text{The sum of internal and external force is equal to zero})$$

or

$$\text{Tibialis anterior force} \times 0.01\,\text{m} = 13.2\,\text{Nm}$$

or

$$\text{Tibialis anterior force} = \frac{13.2\text{Nm}}{0.01\text{m}}$$

or

$$\text{Tibialis anterior force} = 1320\,\text{N}$$

In simple words, we can say that to counter balance an external force of 13.2 Nm at ankle, an internal force of 1320 N needs to be generated by the ankle dorsiflexors as per the hypothetical situation given here. The calculation here is just for understanding and doesn't show the exact value of muscle and GRFV.

20.6 Summary

The calculation of moment, muscle, and joint forces gives us an idea of understanding the segmental biomechanics and application of Newton's laws in more depth. The calculation done by motion analysis software is accurate as they can directly determine the joint centers with the use of markers. In manual calculation, anthropometric data is required to calculate the center of gravity and center of mass for further calculation of joint forces and moments. An ideal static and dynamic equilibrium, the muscles create an internal moment against the external moment of gravity.

References

1. Richards J. Biomechanics in clinic and research. London: Churchill Livingstone; 2008.
2. Richards J. The comprehensive textbook of biomechanics [no access to course] E-Book: [formerly Biomechanics in clinic and research]. Elsevier; 2018.
3. Dempster WT. The anthropometry of body action. 1955.

Correction to: Conceptual Biomechanics and Kinesiology

Correction to: A. Hazari et al Conceptual Biomechanics and Kinesiology, https://doi.org/10.1007/978-981-16-4991-2

The book was inadvertently published with an incorrect author affiliation of Arun G. Maiya as "Department of Physiotherapy, MCHP, MAHE, Manipal, India" whereas it should be "Dean, Manipal College of Health Professions Manipal Academy of Higher Education, Manipal, India".

The updated online version of this book can be found at https://doi.org/10.1007/978-981-16-4991-2

A. Hazari et al., *Conceptual Biomechanics and Kinesiology*,
https://doi.org/10.1007/978-981-16-4991-2_21

GPSR Compliance

The European Union's (EU) General Product Safety Regulation (GPSR) is a set of rules that requires consumer products to be safe and our obligations to ensure this.

If you have any concerns about our products, you can contact us on ProductSafety@springernature.com

In case Publisher is established outside the EU, the EU authorized representative is:

Springer Nature Customer Service Center GmbH
Europaplatz 3
69115 Heidelberg, Germany

Batch number: 10726135

Printed by Printforce, the Netherlands